AF877004

Springer Proceedings in Business and Economics

For further volumes:
http://www.springer.com/series/11960

José-Luis Hervás-Oliver • Marta Peris-Ortiz
Editors

Management Innovation

Antecedents, Complementarities and Performance Consequences

Springer

Editors
José-Luis Hervás-Oliver
Department of Business Organization
Polytechnic University of Valencia
Valencia, Spain

Marta Peris-Ortiz
Department of Business Organization
Polytechnic University of Valencia
Valencia, Spain

ISBN 978-3-319-03133-0 ISBN 978-3-319-03134-7 (eBook)
DOI 10.1007/978-3-319-03134-7
Springer Cham Heidelberg New York Dordrecht London

Library of Congress Control Number: 2013957936

© Springer International Publishing Switzerland 2014
This work is subject to copyright. All rights are reserved by the Publisher, whether the whole or part of the material is concerned, specifically the rights of translation, reprinting, reuse of illustrations, recitation, broadcasting, reproduction on microfilms or in any other physical way, and transmission or information storage and retrieval, electronic adaptation, computer software, or by similar or dissimilar methodology now known or hereafter developed. Exempted from this legal reservation are brief excerpts in connection with reviews or scholarly analysis or material supplied specifically for the purpose of being entered and executed on a computer system, for exclusive use by the purchaser of the work. Duplication of this publication or parts thereof is permitted only under the provisions of the Copyright Law of the Publisher's location, in its current version, and permission for use must always be obtained from Springer. Permissions for use may be obtained through RightsLink at the Copyright Clearance Center. Violations are liable to prosecution under the respective Copyright Law.
The use of general descriptive names, registered names, trademarks, service marks, etc. in this publication does not imply, even in the absence of a specific statement, that such names are exempt from the relevant protective laws and regulations and therefore free for general use.
While the advice and information in this book are believed to be true and accurate at the date of publication, neither the authors nor the editors nor the publisher can accept any legal responsibility for any errors or omissions that may be made. The publisher makes no warranty, express or implied, with respect to the material contained herein.

Printed on acid-free paper

Springer is part of Springer Science+Business Media (www.springer.com)

Foreword

Innovation in the widest sense, arguably, is the only thing that drives organizations and economies forward – everything else is merely intended to keep up with others. As such innovation ought to be a prime concern of decision-makers, whether in the private or public spheres, but equally of those academics who study organizations. Typically, as scholars we are interested under what conditions and how innovation comes about and what consequences it has for a variety of outcome variables.

One form of innovation of particular significance to organizations is innovation in the (internal) practices they use to manage their organizations, what is usually referred to as management innovation, but alternatively as managerial or organizational innovation. In recent years, the topic of management innovation has deservedly received renewed attention from academics in business schools and related departments. Management innovation continues to be a source of competitive advantage, as evidenced by firms that have on the ascendancy in recent years, such as Google and Samsung whose successes can partly be attributed to their management practices.

The community of academics studying various aspects of management innovation is equally growing, and jointly we are producing a range of interesting and exciting new insights. In early 2013, José-Luis Hervas-Oliver, Marta Peris-Ortiz and Francisca Sempere-Ripoll took the initiative to organize a workshop around this theme of management innovation in the city of Valencia, bringing together academics from a number of countries whose work addresses different aspects of management innovation. I was very pleased to be part of this workshop and gathered various new insights from listening to presentations and reading some of the papers. The academics present all demonstrated a deep understanding of management innovation, although each of them approached this phenomenon from a slightly different angle.

This volume presents a collection of the best work presented during the workshop. As such I believe it has a range of new insights to offer to the academic community as well as new practitioners. The chapters in this book address a variety of aspects to do with management innovation, including discussions around how to make sense of this phenomenon and where its boundaries may be, investigations that attempt to disentangle the complex interrelationships between management innovation and other types of innovation such as product or process innovation, work that seeks to shed light on specific aspects of creation and implementation of management innovations through in-depth qualitative methods and work that addresses the performance consequences of management innovation.

Together the chapters in this book address a variety of types of organization, multiple industries, and different facets of management innovation – from structural innovation to people-focused innovation. The book also embraces a variety of methodological approaches, from quantitative modeling using large scale databases through to case studies and grounded theory approaches. In my mind such variety is generally desirable in research on organizations, but is absolutely crucial to improving our understanding of this particular phenomenon. Quantitative testing allows us to observe general patterns of use of management innovation. The Community Innovation Survey for instance offers measures of the implementation of some types of management innovation among a very large number of firms across and beyond Europe and still offers a lot of further potential for investigations into management innovation, as amply demonstrated by various contributors to this volume. By contrast, the more qualitative approaches generate insights into underlying processes and help us understand in much more depth to what extent and in what way context influences creation and implementation (adoption) of management innovation.

The audience for this book will include academic researchers, but I can also see it generate new insights for business and management students and practitioners. This brings me to an important point. Our teaching of business and management to students and executive education generally reflects what is received wisdom, i.e. concepts, cases, and other knowledge important enough to have made it into the 'textbook understanding' of business and management. As my colleague Julian Birkinshaw and I have remarked elsewhere that textbook understanding actually includes prior accumulation of management innovations to a significant degree – such management innovations as multidivisional organizations, total quality management, activity-based costing, return on investment and market segmentation – have very much become part of the received wisdom on how to run organizations effectively. We no longer see them as innovations.

But there is another, related aspect of teaching management innovation that most business schools have struggled with a lot, which is how our knowledge of management innovation can be helpful for generating a stronger capacity to undertake management innovation in real organizations. We do not seem to be good at equipping students with an ability to innovate, preferring to teach what we know about management over teaching students how to innovate in management. Academics tend to shy away from predicting the future, because it is difficult to do so in a reliable way, and often have a hard enough time to keep up with the present – every year my students are surprised that business schools might continue to use case studies that are ten or twenty years old.

In recent years, initiatives such as the Management Innovation Exchange (MIX) (managementexchange.com), started by Gary Hamel and some colleagues, have arisen to tackle some of that deficit. The MIX does this by bringing together a community of enthusiastic practitioners who all share a desire to become more effective management innovators and to see more management innovation in their own organizations. But business schools themselves have yet to make much progress in that respect. And as many have noted, business school academics normally fail to directly engage with organizations to help solve their innovation issues – in that sense engineering and medical schools are much more advanced. I think this book can potentially contribute to such a development towards management innovation being a joint effort between academics and practitioners.

I can therefore only be grateful to the book's editors for allowing me to have been part of this journey of discovery through the workshop and now the publication of this book and want to commend the various authors on their specific contributions, which taken together imply another significant step forward in the study of management innovation. I am sure the reader will enjoy this book and will take away new insights from it.

Coventry, UK Michael J. Mol

Contents

Contributors

Joaquín Alegre Faculty of Economics, Business Department, University of Valencia, Valencia, Spain

Deepa Aravind College of Staten Island – School of Business, City University of New York, Staten Island, NY, USA

Francisco Balbastre-Benavent Universitat de València, Valencia, Spain

Rafael Boix Department of Applied Economics II – Economic Structure, Universitat de València, Valencia, Spain

María Teresa Canet-Giner Universitat de València, Valencia, Spain

Fariborz Damanpour Department of Management and Global Business, Rutgers University, Newark, NJ, USA

Blanca de-Miguel-Molina Department of Business Organization, Universitat Politècnica de València, Valencia, Spain

María de-Miguel-Molina Department of Business Organization, Universitat Politècnica de València, Valencia, Spain

Carlos Devece Department of Business Organization, Universitat Politècnica de València, Valencia, Spain

Naiara Escribá-Carda Universitat de València, Valencia, Spain

Anabel Fernández-Mesa INGENIO (CSIC, Universitat Politècnica de València), Valencia, Spain

Faculty of Economics, Business Department, University of Valencia, Valencia, Spain

Ana García-Granero INGENIO (CSIC, Universitat Politècnica de València), Valencia, Spain

Faculty of Economics, Business Department, University of Valencia, Valencia, Spain

Cesar Gonzalez Instituto Ciencias del Patrimonio (CSIC), Santiago de Compostela, Spain

Charles Heckscher Center for Organizational Learning & Transformation (COLT), Rutgers University, New Jersey, USA

José-Luis Hervás-Oliver Department of Business Organization, Universitat Politècnica de València, Valencia, Spain

Oscar Llopis INGENIO (CSIC, Universitat Politècnica de València), Valencia, Spain

Faculty of Economics, Business Department, University of Valencia, Valencia, Spain

Carolina López-Nicolás Departamento de Organización de Empresas y Finanzas, Campus de Espinardo, Universidad de Murcia, Murcia, Spain

Maria Larraza Malkorra Faculty of Social and Human Sciences, University of Deusto, San Sebastian, Spain

Carlos Martin-Rios Center for Organizational Learning & Transformation (COLT), Rutgers University, New Jersey, USA

Angel Luís Meroño-Cerdán Departamento de Organización de Empresas y Finanzas, Campus de Espinardo, Universidad de Murcia, Murcia, Spain

Marta Peris-Ortiz Department of Business Organization, Universitat Politècnica de València, Valencia, Spain

Gloria Sánchez-González Department of Business Administration, University of León, León, Spain

Francisca Sempere-Ripoll Department of Business Organization, Universitat Politècnica de València, Valencia, Spain

Chapter 1
Management Innovation and Technological Innovation: Friends or Foes?

Marta Peris-Ortiz and José-Luis Hervás-Oliver

Abstract Understanding of management innovation has been advanced in the last decade, but it is still a relatively under-researched topic, at least in comparison with that of technological innovation. This article introduces this volume on Management Innovation; it reviews critically, the multiple conceptual approaches to the topic, looks at the different research streams related to it, and considers the performance consequences or its occurrence. In the latter respect, the article analyzes the synergistic effects of co-adopting management and technological innovation. It also provides a robust theoretical foundation for addressing co-adoption, using a cross-disciplinary perspective. The article also notes that the literature on joint adoption has three blind spots: (i) the literature is fragmented into different, albeit complementary, frameworks and perspectives; (ii) the literature has mainly focused on technological performance, or other general performance, effects deriving from the introduction of management innovations, giving less attention to specific management innovation effects; and (iii) the literature so far has not looked at the joint adoption of specific pairs of technological and management innovations. Finally, as a general point, the article observes that it is surprising how little empirical research has so far gone into exploring the association between the adoption of management innovation and its performance outcomes.

1.1 Introduction

This article addresses the topic of management innovation, looking at its different research streams, and carrying out an in-depth analysis of its performance effects. More specifically, the article carries out a critical review of the management

M. Peris-Ortiz (✉) • J.-L. Hervás-Oliver
Department of Business Organization, Universitat Politècnica de València, Valencia, Spain
e-mail: mperis@doe.upv.es; jose.hervas@omp.upv.es

J.-L. Hervás-Oliver and M. Peris-Ortiz (eds.), *Management Innovation*,
Springer Proceedings in Business and Economics, DOI 10.1007/978-3-319-03134-7_1,

© Springer International Publishing Switzerland 2014

innovation concept, draws up a taxonomy of its different on-going research streams, and theoretically integrates perspectives on the synergistic performance effects of the joint adoption of management and technological innovations. A theoretical integration is achieved by: (i) integrating diverse, but complementary, managerial perspectives addressing the extra benefits of co-adoption, and (ii) dissecting a specific pair of technological and non-technological co-adoption: technological process innovation and organizational innovation. In doing so, this article inserts the topic of management innovation into the mainstream of a diverse set of literatures, and contributes to theory building. This should be of particular value to scholars concerned with understanding the emergent topic of management innovation, and its broad impact on firms.

The article is organized as follows. After this introduction, Section two reviews the management innovation concept; Section three then presents a literature review of different research streams addressing management innovation; Section four describes the extra benefits, or synergistic effects, to be obtained from the co-adoption of management and technological innovation; Section five describes a specific type of joint adoption: technological process innovations together with organizational innovations; finally, in Section six, conclusions and implications are discussed.

1.2 Conceptualizations: A Review

The term *organizational innovation* (Trist and Bamforth 1951) or management innovation (Birkinshaw et al. 2008), encompasses the introduction of new administrative (e.g. Kimberly and Evanisko 1981), organizational (e.g. Armbruster et al. 2008) and managerial (e.g. Birkinshaw et al. 2008) activities. According to Wengel et al. (2000), there are two different kinds of organizational innovation, usually inter-related: *structural* innovations (those that change an organizational arrangement and the division of labour within it), and *managerial* innovations (those relating to the way a firm organizes its activities or its personnel). The notion of management innovation, as distinguished from the technological kind, is rooted in Schumpeter's (1934) concepts of non-technical innovation (such as the opening up of new markets, the development of new sources of supply, and the creation of new market structures). A similar concept is that of administrative or *social innovation* (Damanpour et al. 1989; Trist and Murray 1993), which is said to refer to strategies not directly related to technical innovation, pertaining to policies of recruitment, the allocation of resources, and the structuring of tasks, authority and rewards (Damanpour and Evan 1984; Evan 1966; Kimberly and Evanisko 1981).

Birkinshaw et al. (2008, p. 829) define management innovation as 'the generation and implementation of a management practice, process, structure, or technique that is *new to the state of the art* and is intended to further organizational goals', while Mol and Birkinshaw (2009), and Ganter and Hecker (2013), employ a definition of management innovation which refers to the introduction of management practices that are *new to the firm* and intended to enhance firm performance.

According to Birkinshaw et al. (2008) seminal work, management innovation is formed by *management practices* (that is, what managers do as part of their job on a daily basis); *management processes* (including, for example, strategic planning and performance assessment); and *organizational structure* tasks (such as dealing with communications and re-structuring). The Oslo Manual (2005) distinguishes, within the category of non-technological innovations, between organizational and marketing innovations. An organizational innovation is defined as *the introduction of new-to-the-firm: business practices (such as new forms of quality management); knowledge management systems; organizational methods for the workplace (including those connected to de-centralization, re-structuring, and communication); and management models for external relations (including in respect of outsourcing, alliance formation,* and *inter-firm cooperation).*

In general, the above definitions are rooted in organization theory, and either address practices and policies, or structures and processes. The former relate to the organizational routines mentioned by Simon (1945, p. 46): that is, "*factors that will determine with what skills, values and knowledge the organization member undertakes his work*". The latter address, as stated by Child (1972, p. 2), the "*formal allocation of work roles and the administrative mechanisms to control and integrate work activities*". For the sake of consensus and clarity, we will follow the suggestion of Damanpour and Aravind (2011, p. 35) and view the definitions of administrative, organizational and management innovation as broadly similar, although we recognize that the nuances are quite important. Mol and Birkinshaw (2009), and also Battisti and Stoneman (2010), bring together in their construct of management innovation in the UK, (based on empirical data from the UK CIS), firms' new management practices, new modes of organization, new marketing and new information strategies. However, while the OECD (2005), Mol and Birkinshaw (2009) and Battisti and Stoneman (2010) include in their conceptualizations of management innovation the introduction of new marketing innovations, Armbruster et al. (2008), Camisón and Villar-López (2012), and Damanpour and Aravind (2011) do not. Also, some works only refer to organizational and marketing innovations (from the Oslo Manual) as non-technological innovations. See Table 1.1 for a compilation of definitions and units of measure of the construct. For example, in Table 1.1 it is observed that the construct comprises from occupational roles or compensation, in the early stages, to marketing and strategy in recent works. All in all, the construct is not perfectly delimited.

Occasionally, the innovation literature uses the term "organizational innovation", regardless of the type of innovative outcome developed or introduced in an organization (including technological and non-technological types). In contrast, Lam (2005) defines *organizational innovation* as a precondition for any kind of innovation in organizations. For her, it is necessary to study the relevant and key organizational characteristics which enhance a firm's ability for innovation (e.g. Hall 1992; Hall 1993; Henderson and Cockburn 1994). There is no doubt that one possible barrier to the development of the construct "organizational innovation" is its own *ambiguity and (the) lack of consensus on the definition of the term* (Lam 2004, pp. 31–32).

Table 1.1 Definitions of the management innovation construct and units of measure

Studies	Definitions
Trist and Bamforth (1951)	Organizational innovations:
	There is no stated definition, but the work addresses firms' social structures (including occupational roles, group organization, people, tasks, compensation issues, skills, and working conditions), rather than technical structures. In general, the authors refer to organizational structure, administrative processes (methods of compensation), and the human resource system. This is the pioneering work
Evan (1966)	Administrative innovation:
	The concept relates to: policies of recruitment, the allocation of resources, and the structuring of tasks, authority and rewards
Evan and Black (1967)	Administrative innovation:
	The reference is to aspects such as: human recruitment, jobs allocation, and definition of goals for personnel. The concept covers, basically, administrative and human systems
Damanpour et al. (1989)	Administrative innovation:
	This refers to new techniques related to an organisation's social system, as defined (see above) in Trist and Bamforth (1951), quoted (1989, p. 588)
OECD (2005)	Non-technological innovation:
	This is defined as organizational innovation, which in turn is described as the implementation of a new organizational method in a firm's business practices, workplace organization or external relations
	For example, in the Spanish CIS data questionnaire, organizational innovations are defined as:
	"...*new business practices in the organization of work or in company procedures (for example, in relation to the management of the supply chain, re-engineering, efficient production, quality management, education or training systems.); new knowledge management systems to improve the use or exchange of information or knowledge within the company, or so as to collect information from outside of the company; new organization methods for workplaces in the company, for the purpose of a better distribution of responsibilities and decision-making (for example, using for the first time a system for distributing responsibilities among employees, or managing working teams, or restructuring departments); new models for managing external relations with other companies or public institutions (for example, creating for the first time alliances, associations, or subcontracting arrangements)..*"
	Marketing innovation:
	A marketing innovation is defined by the OECD as the implementation of a new marketing method involving significant changes in product design or packaging, product placement, product promotion, or pricing
	For example, in the Spanish CIS data questionnaire, marketing innovations are defined as:
	"... *significant modifications in the design of the product or in the packaging of the goods or services;*....(This definition excludes changes that affect the functionality of a product or the characteristics of the user. The said changes in the functionality of the product would in fact be considered to be product innovation)...... *new techniques or channels for the promotion of the product;..* (For example, use for the first time of a new advertising channel, or the employment of new trademarks).... *new methods for the positioning of the product in the market or sales channels;* (For example, use for the first time of franchises or distribution licences, or the introduction of direct sale techniques, or the making of exclusive retail agreements) *and, methods for establishing the prices of goods or services.* (For example, use for the first time of a system where prices vary according to demand, or the introduction of price discounts)"

(continued)

Table 1.1 (continued)

Studies	Definitions
Birkinshaw et al. (2008)	Management innovation: This is defined as: 'the generation and implementation of a management practice, process, structure, or technique that is *new to the state of the art* and is intended to further organizational goals'. More specifically, this definition refers to novel or disruptive managerial innovation (such as the M-form invention)
Armbruster et al. (2008)	Organizational innovation: No single definition is provided. The authors carry out a review of studies that have attempted to measure the concept, and find the following terms variously referred to: "team work"; "task integration"; "decentralisation"; continuous improvement processes"; "segmentation of production"; "changes in structures and processes of an organization due to new managerial and working practices such as the introduction of teamwork in production or new supply chain management"; "delegation of responsibility"; "cross-occupational working groups"; "quality circles"; "integration of functions"; and "job rotation"
Battisti and Stoneman (2010)	Organizational innovation: The authors refer to: new management practices, new organization, new marketing and new corporate strategies (They draw on CIS data from UK)
Damanpour and Aravind (2011)	Managerial innovation: The concept relates to management functions which change strategies, structures, systems and administrative procedures
Vaccaro et al. (2012)	Management innovation: The definition is one of new managerial processes, practices, or structures that change the nature of managerial work (such as rules and procedures, employee's tasks and functions, management systems, compensation systems, and communication structures) The authors' identify management innovation by offering respondents the following questions: *1. Rules and procedures within our organization are regularly renewed?* *2. We regularly make changes to our employees' tasks and functions?* *3. Our organization regularly implements new management systems?* *4. The policy with regard to compensation has been changed in the last three years?* *5. The intra- and inter-departmental communication structure within our organization is regularly restructured?* *6. We continuously alter certain elements of the organizational structure?* All responses were measured on a 7-point scale, for which 1 indicated 'strongly disagree' and 7 'strongly agree'
Mol and Birkinshaw (2009) and Ganter and Hecker (2013)	Management innovation: For these authors, management innovation refers to the introduction of management practices that are *new to the firm* and which are intended to enhance firm performance. This definition is based on that of Birkinshaw et al. (2008), drawing on CIS data from the UK. Ganter and Hecker followed German CIS based on Oslo Manual. Mol and Birkinshaw included, from the UK CIS questionnaire: (a) Implementation of new or significantly changed corporate strategies e.g. mission statement, market share, (b) Implementation of advanced management techniques within your firm e.g. knowledge management, quality circles, (c) Implementation of new or significantly changed organizational structures e.g. Investors in people, diversification, and (d) Changing significantly your firm's marketing concepts/strategies e.g. marketing methods

Source: Own

As Larraza (2013, p. 184) states, it is crucial to distinguish clearly between organisational innovation and organisational change. She pointed out that in respect of the Oslo Manual, "*As important for its framework as the standard definition, are also the two characteristics that the Oslo Manual* (OECD 2005) *attributed to organizational innovation: the novelty of the organizational method implemented and the strategic reasons for its deployment. These two features help to differentiate organizational innovation from mere organizational change. Thus, for an organizational change to be considered organizational innovation, it must be completely new to the organization. Furthermore, the mere formulation of management strategies in a document cannot be considered organizational innovation, and its implementation on the firm's activity is a basic requirement. More recent studies have introduced new criteria of differentiation, specifying that the strategic motivation is needed to be considered innovation, orienting it to a considerable improvement of competitive advantage and economic performance for the organization* (Som et al. 2012). *However, this differentiation keeps being confusing since organizational and management literature also includes definitions and empirical research that shows strategic motivation on organizational change processes* (Poole and Van de Ven 2004; Van de Ven 1992)".

1.3 What Do We Know So Far?

1.3.1 Taxonomies

Within the management literature addressing management innovation there are different research streams, but mostly there has been a focus on conceptualizing the phenomenon (Birkinshaw et al. 2008), and understanding its antecedents (e.g. Vaccaro et al. 2012). However, despite an increase in recent years in the number of studies addressing management innovation (e.g. Vaccaro et al. 2012), the topic is still under-researched, at least in comparison to the well-researched phenomenon of technological innovation. As a matter of fact, Crossan and Apaydin (2010) found that out of 524 articles published about innovation in organizations in leading management journals over the period 1981–2008, only three were about management innovation, the majority of papers being classified as addressing technological innovations. Similarly, Keupp et al. (2012) show that of 342 articles reviewed for the period covering 1992–2010, only seven concerned organizational innovations.

In our view, studies of the introduction of management innovation follow three main approaches. Firstly, there are works relating to taxonomies, definitions and the theoretical foundations of the construct, and its systemic organizational implications for innovation in organizations (e.g. Birkinshaw et al. 2008; Damanpour 1991; Evan 1966; Hamel 2006; Wolfe 1994). Secondly, there are those works related to the drivers or antecedents of the adoption of management innovation (Damanpour and Evan 1984; Damanpour 1987; Kimberly and Evanisko 1981; Mol and

Birkinshaw 2009; Wolfe 1994). Thirdly, there are studies focused on the performance consequences of management innovation adoption, including consideration of the synergistic effects of, and the (extra) profit generated by, the joint adoption of more than one type of innovation (Camison-Zornoza and Villar-López 2012; Hervas-Oliver et al. 2012; Mol and Birkinshaw 2009) .

In fact, the phenomenon of joint adoption has been studied through a diverse set of perspectives, and been given various names. For instance, in the technology strategy and innovation literature, the phenomenon has been studied under the names of *synchronous adoption* (Ettlie 1988), and *organizational integration* (Ettlie and Reza 1992), where the concern is to address the optimization of jointly adopted social-oriented and technical-oriented practices (Cua et al. 2001; Damanpour et al. 2009). Also, in the "socio-technical" perspective (Trist and Bamforth 1951), the focus is on the effects of technical (that is the production system oriented to an organization's primary work activity) and social (the human and administrative systems that shape and support the technical one) systems, and the advantages of them being jointly adopted.

There is no doubt that the introduction of new management practices constitutes a crucial tool for leveraging innovation (Birkinshaw et al. 2008; Birkinshaw and Mol 2006; Hamel 2006; Mol and Birkinshaw 2009; Vaccaro et al. 2012). Surprisingly, little research has sought to explain the association between the adoption of organizational innovation and its performance consequences (Birkinshaw and Mol 2006), with some exceptions (e.g. Mol and Birkinshaw 2009).

In respect of studies that address the performance consequences of adopting a management innovation (e.g. Walker et al. 2011), the literature basically divides into two types of approaches. A first group follows a lead-lag approach, that is, one where one innovation mode is seen as a precondition for another, and where, in general, organizational (or management) innovation is seen as necessary before technological innovation adoption (e.g. Damanpour and Evan 1984; Damanpour et al. 1989; Gallego et al. 2012). For example, Gallego et al. (2012), adopt a lead-lag approach in which organizational adoption is considered as a precondition, enabler and facilitator for technological innovation performance. They use a sequential method, although employing cross-sectional data.

A second group focuses on the co-adoption of organizational and technological modes of innovation and its impact on performance. For example, Evangelista and Vezzani (2010) state there is a need for firms to co-adopt technical and non-technical modes of innovation simultaneously, and they measure impact on performance in terms of traditional technical criteria (such as by using sales and productivity variables). Similarly, Battisti and Stoneman (2010) also explore the effects of the synergistic combination of technological and management innovations, and look at whether the introduction of new products and processes increase value added. Neither Evangelista and Vezzani (2010), nor Gallego et al. (2012) nor Battisti and Stoneman (2010) evaluate performance in terms of specific management innovation criteria, but, rather, use technological performance measures. This constitutes an area for improvement. Already, the Oslo Manual provides performance scales aimed at avoiding the sole use of technological performance measures when evaluating the introduction of management innovations.

A consideration of the extra business performance to be accrued from co-adoption have hitherto been limited to: an understanding of the effects of introducing new management innovations on the technological ability to improve product, process and firm performance (measured in terms of better economic, financial, and sales performance) (Camison-Zornoza and Villar-López 2012); on the probability to engage in product or process(e.g. Evangelista and Vezzani 2010) innovations (Gallego et al. 2012); and, on productivity, as measured by sales (Evangelista and Vezzani 2010) or the valued added (Battisti and Stoneman 2010). However, not much is known about the joint adoption of technological and organizational innovations, and its synergistic effects on non-technological organizational-related innovative performance.

1.3.2 About Performance

Why is CIS data appropriate for measuring the implementation of organizational innovation (OI), and its effects? In fact, since 2005, the Oslo Manual (OECD 2005) has incorporated questions about management innovation, its adoption and its effects. In the CIS questionnaire, the Spanish version begins by explaining what is meant by organizational innovation: "*An organizational (management) innovation consists of the implementation of new organizational methods in the internal functioning of the company (including knowledge management methods or systems), in the organization of the workplace, or in respect of external relations, that have not previously been used by the company. It must be the result of strategic decisions made by the management of the company. It excludes mergers or acquisitions, although they may imply an organizational innovation for the Company*".

Then, the questionnaire asks about the introduction of OI: *During the period... did your company introduce* OI? This question (Question 1) asks what specific form of OI has been adopted. Options include: *New business practices in the organization of work or in company procedures* (for example, in relation to the management of the supply chain, re-engineering, efficient production, quality management, education or training systems.); *new knowledge management systems to improve the use or exchange of information or knowledge within the company, or so as to collect information from outside of the company; new organization methods for workplaces in the company, for the purpose of a better distribution of responsibilities and decision-making* (for example, using for the first time a system for distributing responsibilities among employees, or for managing working teams, or engaging in the restructuring of departments); *new models for managing external relations with other companies or public institutions* (for example, creating for the first time alliances, associations, or subcontracting arrangements).

Question 1 has two main uses: indicating whether a firm is an organizational innovator (dummy variable), and indicating the *breadth* of the organizational change by capturing the number of specific types of organizational innovations implemented.

Then, a second question, (Question 2) considers the organizational (management) innovation effects or performance, asking respondents to: "*Indicate the degree of importance of the effects of the organizational innovations introduced by the company during the 2004–2006 period on the following dimensions: reduction of the response period as per the needs of a client or supplier; better quality of goods and services; lower cost per unit produced; improvement in the satisfaction of staff, or decrease in the rotation rates of the same; improvement in the exchange of information, or in communication within the firm*. Respondents are asked to score on a scale of 0–3 (0 none; 1 *low*, 2 *medium*, and 3 *high*)".[1]

In short, it is clear there are alternatives to the use of just technological indicators when measuring management innovation effects.

1.4 Combining Technical and Management Innovation

1.4.1 Fundamentals of Joint Adoption Effects

In order to optimize organizational outcomes, the technical system of an organization should be harmonized with changes in the administrative system (Cummings and Srivastva 1977; Damanpour and Evan 1984; Damanpour et al. 2009; Roberts and Amit 2003; Trist et al. 1993). In this vein, the management literature demonstrates that the successful introduction of new technological activities in industries depends on making simultaneous changes to the organizational structure and to administrative practices (e.g. Thompson 1967). Empirically, it has been observed that firms undertaking management innovation have carried it out in tandem with the carrying out of technological innovation. For instance, Battisti and Stoneman (2010) reported that firms that had introduced innovations in management, organization, strategy and marketing also made changes to products and manufacturing processes. Similarly, in Germany, about half of innovators adopted simultaneously both technological and management alterations; about a third conducted solely management innovations; and around a fifth performed solely technical innovations (Schmidt and Rammer 2007).

Within the economics, strategy and organization, and innovation theoretical perspectives, there has been a consistent tradition of pointing out the complementary advantages that can be achieved in firms by carrying out management and

[1] It is important to notice that the Spanish CIS questionnaire for 2006 and all the previous ones, included that question as EFFECTS. Nevertheless, since 2008, the Spanish questionnaire was modified and changed the variable in order to capture the idea of *objectives* (similar to "innovation goals", related to technological trajectories in the sense of Dosi, 1982) or *factors* for the decision to innovate. Finally, it is important to notice that, although the CIS is standardized for Europe, each country has some peculiarities. For instance, see Spanish questionnaires here: http://www.ine.es/en/daco/daco42/daco4221/ite_cues_en.htm.

technological innovations in tandem[2]. The *innovation literature* (e.g. Ettlie 1988), on the one hand, has traditionally advocated synchronous adoption, although with no strong theoretical foundation. Organization theories, on the other hand, have deeply analyzed the foundations of these complementarities, especially emphasizing the interrelatedness of the way social and technological subsystems function in a firm. For instance, a socio-technical system perspective is used to address organizations (Trist and Bamforth 1951), covering both the social and technological sides of the firm. Lastly, the idea within the economics perspective of *complementarities* (Milgrom and Roberts 1995), highlights the extra profits to be made from joint adoption.

1.4.2 Perspectives

Within the economics field, *complementarities* are said to be achieved in firms by their adoption of management and technological innovations in tandem. Milgrom and Roberts (1995, p. 81) talked about "complements" in a broader sense, as a *relation among groups of activities*, stating that "*…if the levels of any subset of activities are increased, then the marginal return to increases in any or all of the remaining activities rises*". Similarly, Ichniowski et al. (1997) state that the existence of complementarity among practices implies that the magnitude of the performance effect of the entire system is larger than the sum of the marginal effects of adopting each practice individually. Milgrom and Roberts (1995), and Ichniowski et al. (1997) focus on the notion of complementarities as systemic changes among organizational practices, thereby building on contingency theory (Donaldson 1996), in the sense that complementarities among technologies require an adequate *fit* with key organizational variables.

Similarly, in the *strategic management* literature, *complementarities* and their key influence on a firm's innovation capabilities are also recognized (e.g. Stieglitz and Heine 2007). The organizational perspective based on the Resource Based View (RBV) (e.g. Barney 1991; Peteraf 1993) has stressed that a firm's unique internal resources and capabilities determine a firm's performance. Barney (1991) pointed out that the RBV referred to all types of assets, organizational processes, knowledge capabilities and other potential sources of advantage. RBV refers to the internal repository of resources and capabilities to explain heterogeneity in performance. Thus, the capabilities enable the development, deployment and integration of diverse assets, thereby forming a complex bundle of resources that underpin and configure repositories of knowledge, which in turn confer competitive advantage. As Ennen and Richter (2010) suggest, therefore, competitive advantage not only results from developing resources but also from the capability to integrate them in a unique way. Thus, establishing "entire systems of mutually reinforcing design

[2] We really thank Dr. Fariboz Damanpour from Rutgers University for clarification and theoretical support in the identification of the literature fields related to the topic.

elements" enhances performance, and due to the complexity achieved imitation is prevented (e.g. Rivkin 2000).

Teece (1986) defines *complementary assets* as those resources that must be jointly used with an innovation in order to exploit it. Companies can appropriate returns from their innovations when they possess complementary resources that help to achieve inimitability. Dierickx and Cool (1989) also highlight the sustainable competitive advantage that follows from the existence of interconnected assets that prevent imitation.

Summarizing, the point is to understand that achieving competitive advantage requires building systems where the number of elements and their interactions creates an inimitable system (Rivkin 2000). That is, the implementation of a technological innovation together with organizational innovation will integrate social-based (organizational) and technological-based (technology) capabilities, forming superior and complex systems which can lead to above-normal returns (Peteraf 1993).

The *socio-technical* system(Trist and Bamforth 1951) perspective complements the above theories, emphasizing that organizations are made up of people and technology (Pasmore et al. 1982; Trist 1978). The social system, on the one hand, refers to people who work in the organization and the relationships among them. The technical system, on the other hand, consists of techniques, procedures or knowledge used by the social system to achieve organizational goals (Trist and Bamforth 1951). The socio-technical systems perspective establishes that the relationship between organizational subsystems is a correlative relationship representing a "coupling of dissimilarities", where each change in a subsystem requires alterations in the other subsystems (Trist and Murray 1993).

Finally, the innovation and operations management literature has also recognized the necessity to couple technology and management changes. It has been found that innovation activities introduced by technological process innovators simultaneously involve organizational and technological changes (Gopalakrishnan and Damanpour 1997; Reichstein and Salter 2006) that are somewhat blurred and difficult to separate (Edquist et al. 2001; Ettlie and Reza 1992; Womack et al. 1990).

Thus, there has been a consistent finding that the successful introduction of new technology in industries depends on parallel changes in organizational structure and administrative practices (Ettlie 1988; Nabseth and Ray 1974; Thompson 1967).

1.4.3 Social Capabilities Related to Management Innovation: Extending the Socio-technical Approach

Within the socio-technical perspective, the term *organizational innovation* (Trist and Bamforth 1951) was coined to describe the successful development within firms in the mining industry of new social practices (related to organizational and human resources management), and how their integration with technical systems maximized potential outcomes. As (Trist and Bamforth 1951) pointed out: "*It seems... that a qualitative change will have to be effected in the general character of the method, so that a social as well as a technological whole can come into existence. Only if this is achieved can the relationships of the cycle work-group be successfully integrated and a new social balance created*" (37). In the above, the idea of a socio-technical system, or socio-technical integration, is well described. Thus, the socio-technical argument posits that the social and technical systems have to gel together into a single and integrated system.

In our view, a socio-technical approach to management innovation should also address the specific social-based capabilities which are connected to a management innovation. Thus, the introduction of management innovations depends on the specific capabilities developed by organizations in the administrative, managerial and human resource functions of a company. Organizational capabilities (e.g. Grant 1996) mainly concern the organizational rules, procedures and values that are involved in the coordination of functional capabilities and the cohesion of the members of the organization, and provide a knowledge base and proper organizational context for ensuring innovation (e.g. Hall 1992, 1993; Henderson and Cockburn 1994). Other things being equal, those firms which invest and develop more intensively their organizational competencies referred to as a social system (including human, strategic, managerial, structural and administrative functions) will have a higher likelihood of developing social-based capabilities in the sense of Trist and Bamforth (1951). Conversely, the development of technological capabilities influences a propensity for technological innovation, but, unlike social capabilities, is not expected to directly influence organizational innovation. Recent literature about the antecedents of the adoption of organizational innovation has addressed the fact that, in general, the education of the workforce, measured as the percentage of employees with a degree, is potentially an important attribute of the firm and represents one of its key innovation resources, to the extent that many organizational innovations require a high level of skills and education (e.g. Chandler 1962; Ichniowski et al. 1997). Thus, the development of social-based capabilities linked to human systems will be positively linked to the propensity to introduce management innovation.

1.5 Technological Process Innovation and Organizational Innovation Concurrence: Extra Business Performance

Despite a recognition of the performance benefits deriving from co-adoption, the literature has not yet provided empirical evidence about which *specific* technological innovation should be jointly adopted with organizational innovation if extra profits are to occur. In fact, despite the recognition of the value of the joint adoption, or integrative, approach (e.g. Evangelista and Vezzani 2010), deeper analysis has suggested that not all technological innovations will exist in synergy with organizational innovation to the benefit of firm performance. In fact, in the technological strategy and innovation literature (Ettlie 1988; Ettlie and Reza 1992), it is reported that the synergy effects of co-adoption come mainly from the integration of technological *process* innovations and organizational innovations.

What specific form does the integration of technological and organizational innovations take? Innovation activities introduced by process innovators simultaneously involve organizational and technological changes (Gopalakrishnan and Damanpour 1997; Reichstein and Salter 2006) that are somewhat blurred and difficult to separate (Edquist et al. 2001; Ettlie and Reza 1992; Womack et al. 1990). Edquist et al. (2001) include two distinct, but related, activities within the category of process innovation: technological process innovation and organizational process innovation. *Technological process innovations* are new elements that are used in the process of production and include investment goods and intermediate goods such as processing machines, industrial robots and IT equipment. O*rganizational process innovations* are new ways to organize business activities and have no technological elements but rather function via the co-ordination of human resources and work practices, as occurs, for example, in just-in-time production, total quality management or lean production.

The systematic overlap of organizational and technological process innovation is also commonly stressed in the operations management literature (e.g. Duguay et al. 1997; White and Ruch 1990), although most of this literature is based on case studies or specific industries (Ettlie 1988; Womack et al. 1990). Ettlie (1988) finds that better performing organizations synchronise the adaptation of administrative policies with the introduction of technology. Fleck (1994) also recognizes the need to adapt management procedures to the new technology being implemented. Also, Voss (1988) explicitly addresses the complementary effects of integrating new technology with organizational aspects in order to successfully adopt new technology for process innovation. Technology represents an opportunity for re-structuring, and actual outcomes will depend on how the new processes deriving from new technology are integrated with the organization (Barley 1986; Cohen and Zysman 1987; Damanpour 1991; Ettlie and Reza 1992). Therefore, it is expected that the co-adoption of new technological processes and organizational innovations will improve performance. Thus, the more process innovations implemented, then the greater the number of organizational innovations required if new technology is to be properly integrated into the organization.

1.6 Conclusions

This article has focused on the concept of management innovation, its different research streams, and the theoretical foundations of the synergistic effects perceived to occur from joint adoption with technological innovations. Our focus on co-adoption has: (i) aimed to integrate diverse managerial perspectives which address the phenomenon of the extra profits that may ensue from co-adoption, and (ii) addressed the analysis of specific pairs of technological and non-technological co-adoption: technological process and organizational innovations. In doing so, this article inserts the management innovation concept into the mainstream of a diverse set of literatures, contributing to theory building, of particular value to scholars concerned with gaining a better understanding of the complementary resources and capabilities formed when technological and management innovations are jointly adopted.

Through this study we have contributed to the management innovation phenomenon in several ways. Firstly, this work has dissected the management innovation concept, and its associated research streams, by identifying and setting out the body of theoretical knowledge connected to it, and by critically reviewing past and current empirical evidence.

Secondly, an integrated technological and organizational innovation framework has been constructed by bringing together, or "cross-fertilizing", diverse, but complementary, managerial and organizational perspectives, which embraces and describes the positive and synergistic advantages of implementing management innovation jointly with technological innovations. As a matter of fact, in the management literature, evidence has already been produced of the positive gain from simultaneous concurrence (Battisti and Stoneman 2010; Damanpour et al. 2009). Similarly, the innovation perspective (e.g. Ettlie 1988), and specifically the *organizational integration* approach (Ettlie and Reza 1992), has also claimed positive results from *complementarities,* in line with those suggested in the economics field of writings (Milgrom and Roberts 1995). Additionally, the occurrence of a congruence of technological and organizational (or social-based) capabilities in firms is also reflected in the literature about socio-technical systems(Trist and Bamforth 1951), which emphasizes that organizations are made up of technological systems and human systems (Pasmore et al. 1982; Trist 1978), which need to be integrated. Then, in the strategic management perspective, the identification of *complementary assets or resources and capabilities* (Teece 1986) also provides a robust theoretical foundation for the perceived synergistic effects of joint adoption.

Thirdly, this article has shown that the specific introduction of technological process innovation frequently occurs concurrently with the introduction of organizational (as a particular type of management) innovation, creating synergistic effects, or complementarities, which are positively related to performance. Therefore, the co-adoption of technological process innovations and organizational innovations is beneficial for a firm's performance, due to the fact that this integrated pair of innovations forms a superior combination of assets, a "coupling of dissimilarities", satisfying the need that each change in a subsystem requires alterations in other subsystems (Trist and Murray 1993).

There is no doubt that the occurrence, or not, of management innovation is crucial for the understanding of firms' strategy and the creation of competitive advantage. This latter task, however, is ours.

Acknowledgment This paper has been financially supported by the Spanish Ministry of Economics (ECO2010-17318) *Innoclusters*. Furthermore, Authors gratefully acknowledge support from the Universitat Politècnica de València through the project Paid-06-12 (Sp 20120792).

References

Armbruster H, Bikfalvi A, Kinkel S, Lay G (2008) Organizational innovation: the challenge of measuring non-technical innovation in large-scale surveys. Technovation 28(10):644–657

Barley SR (1986) Technology as an occasion for structuring: evidence from observations of CT scanners and the social order of radiology departments. Adm Sci Q 31:78–108

Barney J (1991) Firm resources and sustained competitive advantage. J Manage 17(1):99–120

Battisti G, Stoneman P (2010) How innovative are UK firms? Evidence from the fourth UK community innovation survey on synergies between technological and organizational innovations. Br J Manage 21(1):187–206

Birkinshaw JM, Mol MJ (2006) How management innovation happens. MIT Sloan Manage Rev 47(4):81–88

Birkinshaw J, Hamel G, Mol M (2008) Management innovation. Acad Manage Rev 33(4): 825–845

Camison-Zornoza C, Villar-López A (2012) On how firms located in an industrial district profit from knowledge spillovers: adoption of an organic structure and innovation capabilities. Br J Manage 23(3):361–382

Chandler AD (1962) Strategy and structure: chapters in the history of the American industrial enterprise. MIT Press, Cambridge

Child J (1972) Organizational structure, environment and performance: the role of strategic choice. Sociology 6(1):1–22

Cohen S, Zysman J (1987) Manufacturing matters: the myth of the post-industrial economy. Basic Books, New York

Crossan M, Apaydin M (2010) A multi-dimensional framework of organizational innovation: a systematic review of the literature. J Manage Stud 47(6):1154–1191

Cua KO, McKone KE, Schroeder RG (2001) Relationships between implementation of TQM, JIT, and TPM and manufacturing performance. J Oper Manage 19(6):675–694

Cummings TG, Srivastva S (1977) Management of work: a socio-technical systems approach. Comparative Administration Research Institute, Kent: distributed by Kent State University Press. Translation of: (Kent, Ohio)

Damanpour F (1987) The adoption of technological, administrative, and ancillary innovations: impact of organizational factors. J Manage 13(4):675

Damanpour F (1991) Organizational innovation – a metaanalysis of effects of determinants and moderators. Acad Manage J 34(3):555–590

Damanpour F, Aravind D (2011) Managerial innovation: conceptions, processes, and antecedents. Manage Organ Rev 8(2):423–454

Damanpour F, Evan WM (1984) Organizational innovation and performance: the problem of organizational lag. Adm Sci Q 29(3):392–409

Damanpour F, Szabat KA, Evan WM (1989) The relationship between types of innovation and organizational performance. J Manage Stud 26(6):587–602

Damanpour F, Walker RM, Avellaneda CN (2009) Combinative effects of innovation types and organizational performance: a longitudinal study of service organizations. J Manage Stud 46(4):650–675

Dierickx I, Cool K (1989) Asset stock accumulation and sustainability of competitive advantage. Manage Sci 35(12):1504–1511

Donaldson L (1996) For positivist organization theory: proving the hard core. Sage, London

Duguay CR, Landry S, Pasin F (1997) From mass production to flexible/agile production. Int J Oper Prod Manage 17(12):1183–1195

Edquist C, Hommen L, McKelvey MD (2001) Innovation and employment: process versus product innovation. European socio-economic research Conference, Edward Elgar Publishing, Brussels

Ennen E, Richter A (2010) The whole is more than the sum of its parts or is it? A review of the empirical literature on complementarities in organizations. J Manage 36(1):207–233

Ettlie JE (1988) Taking charge of manufacturing: how companies are combining technological and organizational innovations to compete. Jossey-Bass, San Francisco, p 643
Ettlie JE, Reza EM (1992) Organizational integration and process innovation. Acad Manage J 35(4):795–827
Evan WM (1966) The organization set: toward a theory of interorganizational relations. In: Thompson JD (ed) Approaches to organizational design. University of Pittsburgh Press, Pittsburgh
Evan WM, Black G (1967) Innovation in business organizations: some factors associated with success or failure of staff proposals. J Bus 40(4):519–530
Evangelista R, Vezzani A (2010) The economic impact of technological and organizational innovations a firm-level analysis. Res Policy 39(10):1253–1263
Fleck J (1994) Learning by trying: the implementation of configurational technology. Res Policy 23(6):637–652
Gallego J, Rubalcaba L, Hipp C (2012) Organizational innovation in small European firms: a multidimensional approach. Int Small Bus J 31(5):563–579
Ganter A, Hecker A (2013) Deciphering antecedents of organizational innovation. J Bus Res 66(5):575–584
Gopalakrishnan S, Damanpour F (1997) A review of innovation research in economics, sociology and technology management. Omega 25(1):15–28
Grant RM (1996) Toward a knowledge-based theory of the firm. Strateg Manage J 17:109–122
Hall R (1992) The strategic analysis of intangible resources. Strateg Manage J 13(2):135–144
Hall R (1993) A framework linking intangible resources and capabilities to sustainable competitive advantage. Strateg Manage J 14(8):607–618
Hamel G (2006) The why, what, and how of management innovation. Harv Bus Rev 84(2):72–84
Henderson R, Cockburn I (1994) Measuring competence? Exploring firm effects in pharmaceutical research. Strateg Manage J 15(S1):63–84
Hervas-Oliver JL, Sempere-Ripoll F, Boronat-Moll C (2012) Process innovation objectives and management complementarities: patterns, drivers, co-adoption and performance effects, UNU-MERIT Working Paper, 2012–51 http://www.merit.unu.edu/publications/wppdf/2012/wp2012-051.pdf. Accesed 8 Oct 2013
Ichniowski C, Shaw K, Prennushi G (1997) The effects of human resource practices on manufacturing performance: a study of steel finishing lines. Am Econ Rev 87(3):291–313
Keupp M, Palmié M, Gassmann O (2012) The strategic management of innovation: a systematic review and paths for future research. Int J Manage Rev 14(4):367–390
Kimberly JR, Evanisko MJ (1981) Organizational innovation: the influence of individual, organizational, and contextual factors on hospital adoption of technological and administrative innovations. Acad Manage J 24:689–713
Lam A (2004) Organizational innovation. Brunel Business School, Brunel
Lam A (2005) Organizational innovation. In: Fagerberg J, Mowery DC, Nelson RR (eds) The Oxford Handbook of Innovation. Oxford University Press, Oxford 115–147
Larraza M (2013) Understanding organizational innovation from its practice. The organizational innovation workshop. February 2013. Valencia
Milgrom P, Roberts J (1995) Complementarities and fit strategy, structure, and organizational change in manufacturing. J Account Econ 19(2):179–208
Mol M, Birkinshaw J (2009) The sources of management innovation: when firms introduce new management practices. J Bus Res 62(12):1269–1280
Nabseth L, Ray GF (1974) The diffusion of new industrial processes: an international study. Cambridge University Press, New York/London
OECD (2005) Oslo manual: guidelines for collecting and interpreting innovation data. OECD, Paris
Pasmore W, Francis C, Haldeman J, Shani A (1982) Sociotechnical systems: a north American reflection on empirical studies of the seventies. Hum Relat 35(12):1179–1204
Peteraf M (1993) The cornerstones of competitive advantage: a resource-based view. Strateg Manag J 14(3):179–191

Poole MS, Van de Ven AH (2004) Central issues in the study of change and innovation. In: Poole MS, Van de Ven AH (eds) Handbook of organizational change and innovation. Oxford University Press, New York, p 3
Reichstein T, Salter A (2006) Investigating the sources of process innovation among UK manufacturing firms. Ind Corp Change 15(4):653–682
Rivkin JW (2000) Imitation of complex strategies. Manage Sci 46(6):824–844
Roberts PW, Amit R (2003) The dynamics of innovative activity and competitive advantage: the case of Australian retail banking, 1981 to 1995. Organ Sci 14(2):107–122
Schmidt T, Rammer C (2007) Non-technological and technological innovation: strange bedfellows? ZEW Discussion Papers 07-052
Schumpeter JA (1934) The theory of economic development: an inquiry into profits, capital, credit, interest, and the business cycle. Transaction Publishers, Cambridge, MA
Simon HA (1945) Administrative behaviour. The Macmillan Company, New York
Som O, Diekmann J, Solberg E, Schricke E, Schubert T, Jung-Erceg P, Stehnken T, Daimer S (2012) Organisational and marketing innovation – promises and pitfalls? PRO INNO Europe: INNO-Grips II report, Brussels: European Commission, DG Enterprise and Industry. http://ec.europa.eu/enterprise/policies/innovation/files/proinno/innovation-intelligence-study-5_en.pdf. Accessed 8 Oct 2013
Stieglitz N, Heine K (2007) Innovations and the role of complementarities in a strategic theory of the firm. Strateg Manage J 28(1):1–15
Teece DJ (1986) Profiting from technological innovation: implications for integration, collaboration, licensing and public policy. Res Policy 15(6):285–305
Thompson JD (1967) Organizations in action. McGraw Hill, New York
Trist EL (1978) On socio-technical systems. In: Sociotechnical systems: a sourcebook. University Associates, San Diego
Trist EL, Bamforth KW (1951) Some social and psychological consequences of the longwall method: an examination of psychological situation and defences of a work group in relation to the social structure and technological content of the work system. Hum Relat 4(3)
Trist E, Murray H (1993) In: Trist E, Murray H (eds) The socio-technical perspective. University of Pennsylvania Press, Pennsylvania
Trist E, Murray H (eds) (1993) The social engagement of social science: a tavistock anthology, vol. II: the socio-technical perspective. University of Pennsylvania Press, Philadelphia, PA
Vaccaro IG, Jansen JJP, Van Den Bosch FAJ, Volberda HW (2012) Management innovation and leadership: the moderating role of organizational size. J Manage Stud 49(1):28–51
Van de Ven AH (1992) Suggestions for studying strategy process: a research note. Strateg Manage J 13(5):169–188
Voss CA (1988) Implementation: a key issue in manufacturing technology: the need for a field of study. Res Policy 17(2):55–63
Walker RM, Damanpour F, Devece CA (2011) Management innovation and organizational performance: the mediating effect of performance management. J Public Adm Res Theory 21(2):367
Wengel J, Lay G, Nylund A, Bager-Sjögren L, Stoneman P, Bellini N, Bonaccorsi A, Shapira P (2000) Analysis of empirical surveys on organisational innovation and lessons for future community innovation surveys. Scientific Follow-Up of the Community Innovation Survey (CIS) Project CIS 98(191), p 483
White RE, Ruch WA (1990) The composition and scope of JIT. Oper Manage Rev 7(3/4):9–18
Wolfe RA (1994) Organizational innovation – review, critique and suggested research directions. J Manage Stud 31(3):405–431
Womack JP, Jones DT, Roos D (1990) The machine that changed the world: the story of lean production. Rawson Associates, New York

Chapter 2
Combining Technological and Management Innovations: Empirical Evidence of a Premium Effect

Francisca Sempere-Ripoll, José-Luis Hervás-Oliver, and Marta Peris-Ortiz

Abstract This paper determines the performance consequences of the joint adoption of organizational and technological innovations. Using CIS data to analyse 2,837 technological innovators, it is empirically demonstrated that a firm's co-adoption strategy leads to a premium technological performance thanks to the extra synergistic effects, or the generation of complementarities. Firms co-adopting outperform those adopting solely technological innovations. Therefore, the combination of technical and non-technical competencies, and the integration of a socio-technical system, creates exceptional innovation capabilities. The strategic adoption of solely technological innovation is not enough: in order to leverage the performance of technological innovation it should be integrated into the organization.

2.1 Introduction

An innovation mode is defined as a *set of typologies of innovation practices, strategies and performances* (Evangelista and Vezzani 2010), i.e., the differing types of behaviour a company displays with regard to investment in innovation activities in order to create new knowledge in different innovative outputs. This has traditionally concerned product and process innovations, which have particular objectives such as the improvement of existing products or a reduction in labour costs, which are "technical goals" (in the sense of Cohen and Malerba 2001, p. 590). There is also the *organizational or management innovation* mode, also known as administrative (Evan 1966) or organizational innovation (OECD 2005), which refers to strategies that are not directly related to technical innovation. While a technical or technological innovation relates to a new product or process, an administrative innovation

F. Sempere-Ripoll (✉) • J.-L. Hervás-Oliver • M. Peris-Ortiz
Department of Business Organization, Universitat Politècnica de València, Valencia, Spain
e-mail: fsempere@omp.upv.es; jose.hervas@omp.upv.es; mperis@doe.upv.es

J.-L. Hervás-Oliver and M. Peris-Ortiz (eds.), *Management Innovation*, Springer Proceedings in Business and Economics, DOI 10.1007/978-3-319-03134-7_2,
© Springer International Publishing Switzerland 2014

pertains to policies of recruitment, the allocation of resources, and the structuring of tasks, authority and rewards (Evan 1966), or even marketing efforts (Mol and Birkinshaw 2009; OECD 2005).

By *management innovation* in this paper we adhere to the two definitions of non-technological innovation provided by the Oslo Manual (OECD 2005), that is, organization and marketing innovations. Organizational innovation is defined as "the implementation of a new organizational method in the firm's business practices, workplace organization or external relations" (OECD 2005, p. 177), stressing the fact that it is the result of strategic decisions taken by management (pp. 51). Similarly, a marketing innovation is defined as the "implementation of a new marketing method involving significant changes in product design or packaging, product placement, product promotion or pricing" (pp. 49). Birkinshaw et al. (2008) offer a similar definition to the Oslo Manual's specification for organizational innovation when they define the term management innovation as involving new organizational structures, administrative systems, and management practices, processes and techniques.

Studies on the adoption of management or non-technical innovation are less abundant (e.g. Birkinshaw et al. 2008; Hamel 2006; Vaccaro et al. 2012; Walker et al. 2011) than those on the well-researched technological innovation. Despite an increase in the number of studies addressing the joint adoption of management and technological innovation (e.g. Battisti and Stoneman 2010; Camison-Zornoza and Villar-López 2012; Evangelista and Vezzani 2010) more empirical evidence is needed. As a matter of fact, there has been a call for "…a greater emphasis to the integration of technological and organizational factors" (Battisti and Stoneman 2010, p. 203), and, similarly, a need for more empirical work that validates the positive returns to be had from co-adoption (Evangelista and Vezzani 2010). The rationale for these demands is that most firms do not engage solely in management innovation, but simultaneously adopt both technological and management modes. In Germany, it has been found that about half of innovators adopt simultaneously both technological and management modes; about a third conduct solely management innovations; and around a fifth perform solely technical innovations (Schmidt and Rammer 2007). This paper focuses on the research gaps and contributes with empirical evidence.

The literature has stressed the fact that gaining technical competencies will require an adaptation of both the technology itself and the organization (Ettlie 1988; Ettlie and Reza 1992; Fleck 1994; Leonard-Barton and Deschamps 1988; Voss 1988). In this paper we propose to investigate the idea of a need for the organizational integration of technology by using a large-scale database, the Community Innovation Survey (CIS) by Eurostat, to answer whether a technology innovation strategy alone is enough to maximize performance, or whether in fact joint adoption is required. Specifically, this paper determines the effects that a joint adoption strategy exerts on a firm's performance, exploring theoretically and empirically the potential synergistic benefits which can result from the joint adoption of technological and management innovation strategies. In particular, we look at the consequences of the introduction of new management practices for technological innovation performance. In other words, this paper shows how technological innovation performance is leveraged when a firm complements it with other non-technological innovations, when, that is to say, the technology is integrated into the organization.

The paper contributes to the literature by establishing a theoretical framework through which to consider the connection between technological innovation and management innovation, and in doing so brings together the literature focussed solely on technological innovation with that concerned with organization theory and strategy.

The paper uses a large dataset (2,837 firms) based on the Spanish CIS data for 2006, enabling a cross-industry generalization of results and for theory building to take place. The paper is organized as follows. In Sect. 2.2 we make a review of the theory before setting out our hypothesis. Section 2.3 outlines the dataset, variables and statistical methods. We then present and discuss the results in Sect. 2.4. Finally, Sect. 2.5 concludes the analysis and discusses briefly the theoretical contributions, together with insights and implications for scholars, managers and policymakers.

2.2 Theoretical Framework

2.2.1 Organizational Integration of Technology Through Joint Adoption

The idea of joint adoption is rooted in the *socio-technical* literature. Thus, the perspective of the socio-technical system (Trist and Bamforth 1951) describes a method of viewing organizations that emphasises the interrelatedness, and necessary integration, of the functioning of the social and technological subsystems, contending that organizations are made up of people and technology (Pasmore et al. 1982; Trist 1978). In socio-technical systems there is a correlative relationship between organizational subsystems, representing a "coupling of dissimilarities" where each subsystem change requires changes in the other subsystems (Trist et al. 1993; Trist and Bamforth 1951).

The socio-technical perspective has been complemented in the technology innovation strategy literature by many case studies and industry-specific studies which have found that management innovation is an effective way of complementing and supporting technical innovation (Ettlie 1988; Ettlie and Reza 1992; Fleck 1994; Leonard-Barton and Deschamps 1988; Voss 1988; Womack et al. 1990). Ettlie (1988) dubs the simultaneous use of management innovation and technological innovation "synchronous innovation", and argues that the resulting ability to realise complementarities and synergies has meant that the use of appropriate forms of management innovation has made technological innovation more effective.

Thus, it is recognized that the sourcing and implementing of a new technology in a firm to gain technical competencies will require an adaptation of both the technology itself and the organization (Ettlie 1988; Ettlie and Reza 1992; Fleck 1994;

Leonard-Barton and Deschamps 1988; Voss 1988). As evidence of this, Luria (1987) shows that changes in organizational structure or technology alone did not yield any significant cost reductions in automobile component plants, due to the fact that co-adoption is needed for stimulating the complementarities between the two distinct but related innovation modes. Therefore, managers should recognize and assume responsibility for both technological and organizational change (Leonard-Barton and Deschamps 1988; Voss 1988), adapting the existing management procedures to the new technology (Fleck 1994), and thus capturing the complementary effects of integrating new technology into the organization (Voss 1988).

The evidence from both the socio-technical and the technology innovation strategy literatures is that integration optimizes organizational outcomes (Cummings and Srivastva 1977; Damanpour and Evan 1984; Damanpour et al. 2009; Roberts and Amit 2003; Trist et al. 1993). In other words, a simple adoption of a technological innovation mode is not sufficient to achieve a sustained competitive advantage: the complementary adoption of administrative or organizational innovations is also required (Bloom and Van Reenen 2007; Hamel 2006; Wengel et al. 2000). This is necessary to fully realize the potential synergistic effects of bringing together the technical and social systems. Therefore, it is expected that the joint adoption of both types of innovation will lead to superior returns, or *a premium effect*, because technological performance will be thereby amplified. Consequently we present the following hypothesis:

Hypothesis: The joint adoption of technological and management innovations has synergistic effects which render better performance.

2.3 Empirical Design

2.3.1 Dataset and Sample

The dataset is the Spanish Community of Innovation Survey (CIS). The method and the types of questions in CIS are described in the Oslo Manual (Oslo Manual: OECD 2005) for 2006. CIS data is used for the study of innovation at the firm level in a large number of studies in Belgium, France, Spain, UK and other countries (Cassiman and Veugelers 2002; Crepon et al. 1998; Escribano et al. 2009; Evangelista and Vezzani 2010; Laursen and Salter 2006; Mol and Birkinshaw 2009; Tether 2002). To see the survey and the details of the questionnaire design, visit Eurostat.[1] CIS data waves cover 3 years, and in the 2006 wave, the questions asked refer to the 2004–2006 period.

In respect of our sample, our empirical analysis was limited to examining all technological innovators available in the Spanish CIS data for 2006, using the industry classification by NACE-93, from 15 to 74 (2-digits, 59 sectors).

[1] To know more, please retrieve information from: http://epp.eurostat.ec.europa.eu/cache/ITY_SDDS/en/inn_esms.htm; http://epp.eurostat.ec.europa.eu/cache/ITY_PUBLIC/OSLO/EN/OSLO-EN.PDF.

The final sample consisted of 2,837 technological innovators, defined as firms that introduced simultaneously new products and processes during the period covered. From this sample, we distinguished those technological innovators which *in addition* introduced management innovations (1,704, 60 % of the sample), and they were named *management innovators*. Thus, management innovators carried out all four types of innovation: technological (both in respect of products and processes) and management innovation (both organizational and marketing). This is a *convenience* sample that enables research on the effects of joint adoption by comparing the performance of a group of technological innovators against a group of management innovators.

Fifty seven percentage of the sample consisted of small firms, i.e. with less than 50 employees, and 32 % were medium-sized companies, having 50–250 employees. Also, 68 % of firms in the sample were exporters, and 28 % belonged to industrial groups. Spain is a technology follower due to its predominantly low-tech industry, and the sample confirmed this fact. Sixty five percentage of the sample was made up of low-tech and medium low-tech industries, and only 13 % were high-tech industries.

2.3.2 Empirics

Despite the limitations of the CIS questionnaire, due to the fact that its initial versions were designed to measure technological innovation (mainly product related ones), the incorporation of non-technological innovations after the 3rd edition (starting in 2006 in the Spanish CIS questionnaire) has presented a great opportunity to explore and shed light on the different innovation strategies, and on joint effects.[2]

[2] The technological innovators are the firms that carried out product and process innovations in tandem. **A product innovator** is the one that has introduced at least one new or significantly improved goods or services during the research period. **A process innovators** is the one that has introduced at least one of the following: new or significantly improved methods for the manufacture or production of goods or services, new or significantly improved logistics systems or delivery or distribution methods for its supplies, goods or services, or support activities for its processes, such as systems of maintenance or IT operations, of purchases or of accounting, being new or significantly improved. We considered the introduction of **management innovation** when the firm carried out organization and marketing innovation in tandem. In this definition of management innovation, we consider both organizational and marketing innovation. A firm is considered **organizational innovator** if introduce at least one of the following innovations: new business practices in the organization of the work or of the company procedures, new knowledge management systems to improve the use or exchange of information, knowledge and ability, within the company, or so as to collect information outside of the company, new organization methods for the workplaces in the company, for the purpose of a better distribution of responsibilities and decision-making, or new management models for external relations with other companies or public institutions?. **And a marketing innovator** may have carried out at least one of the following: significant modifications in the design of the product or in the packaging of the goods or services, new techniques or channels for the promotion of the product, new methods for the positioning of the product in the market or sales channels, or new methods for establishing the prices of the goods or services.

In respect of a performance variable, we looked at the technological innovation effects of adopting both product and process technological innovation. This measure is appropriate due to the fact that the whole sample is composed of technological innovators (of which, some were also engaged in management innovation, and some not), and all the companies in it answered the appropriate questions from the Spanish CIS questionnaire. In the Spanish CIS 2006 version, the question defining our dependent variable in respect of the effects of technological innovation during the research period is as follows: "*The result of the innovative activity may have had different effects on the company. Indicate the degree of importance of the following effects: wider range of product or services; increase market share; higher quality of products or services; higher production flexibility; higher production capacity; lower labour cost per unit; fewer materials and energy per produced unit; and lower environmental impact and compliance with environmental standards*". Effects were measured on a four-point scale: No effect=0; Low effect=1; Medium effect=2; High effect=3.

2.4 Results

We compared two groups, management innovators and the remaining technological innovators (referred to below as non-management innovators), using an *Anova* test, in order to understand the additional effects of management innovation. The dependent variable for the *Anova* test was based on the technological innovation effects described above. In Table 2.1 it is observed how, systematically and consistently, all technological innovation effects are statistically significant at $p<0.01$, showing that the management innovators outperform non-management innovators. In short, it can be said that the joint adopters (the management innovators, that is, those performing both management and technological innovations) obtain synergistic effects in their technological performance by integrating technology into the organization and thus generating extra business benefits. Therefore, joint adoption increases returns. See Table 2.1.

As the results show, the introduction of management innovations, as composed of organization and marketing innovations, complement and reinforce technological innovation performance, due to the complementary or synergistic effects of joint adoption. It is empirically demonstrated that the use of a more systemic approach to innovation by the adoption of technological and management modes in tandem leads to superior performance, confirming the views expressed in the literature (e.g. Battisti and Stoneman 2010; Bloom and Van Reenen 2006; Ruigrok et al. 1999). However, this result does not mean that there is a cause-and-effect relationship between the two, but, rather, a positive synergistic gain, and this finding has been supported by the literature. For example, Damanpour et al. (2009, p. 658) state that the relationship between the technical and the non-technical systems in social-technical systems theory is a correlative relationship, presenting a "coupling of dissimilarities" (Damanpour and Evan 1984; Scott 1992). In summary, concentrating solely on either the technical or the non-technical will result in a poorer performance level, as previously stated by, for example, Herbst (1974).

Table 2.1 Anova analysis comparing management innovators with non-management innovators

Type of effects	Group of firms	N	Mean	Standard Deviation	F	Significance
Wider range of product or services	Non-management innovators	1,137	1.99	1.02	94.98	0.00*
	Management innovators	1,704	2.33	0.87		
Increase market share	Non-management innovators	1,137	1.71	1.05	141.85	0.00*
	Management innovators	1,704	2.13	0.90		
Higher quality of products or services	Non-management innovators	1,137	2.09	0.98	109.79	0.00*
	Management innovators	1,704	2.43	0.80		
Higher production flexibility (product or service)	Non-management innovators	1,137	1.74	0.98	90.12	0.00*
	Management innovators	1,704	2.06	0.86		
Higher production capacity	Non-management innovators	1,137	1.83	1.01	66.55	0.00*
	Management innovators	1,704	2.12	0.89		
Lower labour cost per unit	Non-management innovators	1,137	1.33	1.03	65.72	0.00*
	Management innovators	1,704	1.62	0.96		
Fewer materials and energy per produced unit	Non-management innovators	1,137	1.12	0.99	44.88	0.00*
	Management innovators	1,704	1.37	1.00		
Lower environmental impact and compliance with environmental standards[a]	Non-management innovators	1,137	1.15	1.13	60.97	0.00*
	Management innovators	1,704	1.48	1.14		

Level of significance: *$p<0.01$

[a]An average of the two effects is used

2.5 Conclusions

This work has focused on analysing the performance consequences of the joint adoption of technological and management innovations. Specifically, this paper has shown empirically the synergistic effects from joint adoption, which result in a premium effect, or extra benefits following the integration of firms' social and technical systems.

By analysing 2,837 firms from 2006 CIS data in Spain, it has been empirically demonstrated that the joint adoption of technological and management innovations renders higher technological returns than does the sole adoption of the technological innovation mode. Thus, the combination of technical and non-technical competencies and capabilities in the innovation process produces exceptional capabilities due to the synergistic benefits from integration (Trist et al. 1993). In other words, the "winning innovation strategy" that firms should adopt in order to improve innovative performance is the simultaneous adoption of technological and management innovations, which enables the integration of the socio-technical system, and which increases a firm's technological performance. This result confirms our hypothesis stated earlier. Put differently, the sole adoption of the technological mode is not enough: in order to leverage the performance consequences of technological innovation it should be coupled into the organization (Bloom and Van Reenen 2007; Hamel 2006). Thus, technology is only one part of the story, and a substantial unexplained productivity differential still remains, which panel data econometricians often label as the fixed effects of "managerial quality" (Mundlak 1961).

The paper has implications for scholars, policymakers and managers. For scholars, the paper represents a first attempt at reinforcing and legitimising the field of research in management innovation by linking the topic to that of technological innovation. For policymakers, the paper's conclusions make clear that policies should include the promotion of adoption by firms of management innovation. Thus, tax breaks and other public initiatives should take this into account, reinforcing and promoting the adoption of management innovations which support and complement technological innovation activities. For managers, it is important they become aware of the benefits of non-technological activities for promoting an organizational context which fosters innovation in a broad sense.

The paper suffers from limitations. Firstly, the paper uses a convenience sample. Secondly, the analysis of large-scale databases like the CIS one used in this study poses many questions that cannot be investigated without other observational research methods. Lastly, the results are limited to Spain and research concerning other countries may produce different results.

Acknowledgments The authors would like to thank the Spanish Ministry of Science and Innovation (currently the Ministry of Economics and Competitiveness) for financially supporting this research (ECO2010-17318 MICINN, "Innoclusters") and authors would also like to thank Universitat Politècnica de València for supporting this research through the proyect Paid-06-12 (Sp 20120792).

Dr. Hervas is also in debt with GVA BEST 2011 program visiting London School of Economics at UK, founded by the Valencia Region Government (GVA). We also thank Dr. Damanpour for its patience and support clarifying some of the theoretical issues presented in this paper. Usual disclaimers apply.

References

Battisti G, Stoneman P (2010) How innovative are UK firms? Evidence from the fourth UK community innovation survey on synergies between technological and organizational innovations. Br J Manage 21(1):187–206

Birkinshaw J, Hamel G, Mol M (2008) Management innovation. Acad Manage Rev 33(4):825–845

Bloom N, Van Reenen J (2006) Measuring and explaining management practices across firms and countries. National Bureau of Economic Research, Cambridge, MA

Bloom N, Van Reenen J (2007) Measuring and explaining management practices across firms and countries. Q J Econ 122(4):1351–1408

Camison-Zornoza C, Villar-López A (2012) On how firms located in an industrial district profit from knowledge spillovers: adoption of an organic structure and innovation capabilities. Br J Manage 23(3):361–382

Cassiman B, Veugelers R (2002) R&D cooperation and spillovers: some empirical evidence from Belgium. Am Econ Rev 92(4):1169–1184

Cohen W, Malerba F (2001) Is the tendency to variation a chief cause of progress? Ind Corp Change 10(3):587–608

Crepon B, Duguet E, Mairesse J (1998) Research, innovation, and productivity: an econometric analysis at the firm level. Econ Innov New Technol 7(2):115–159

Cummings TG, Srivastva S (1977) Management of work: a socio-technical systems approach. Comparative Administration Research Institute, distributed by Kent State University Press, Kent, Translation of: (Kent, Ohio)

Damanpour F, Evan WM (1984) Organizational innovation and performance: the problem of organizational lag. Adm Sci Q 29(3):392–409

Damanpour F, Walker RM, Avellaneda CN (2009) Combinative effects of innovation types and organizational performance: a longitudinal study of service organizations. J Manage Stud 46(4):650–675

Escribano A, Fosfuri A, Tribo JA (2009) Managing external knowledge flows: the moderating role of absorptive capacity. Res Policy 38(1):96–105

Ettlie JE (1988) Taking charge of manufacturing: how companies are combining technological and organizational innovations to compete. Jossey-Bass, San Francisco, p 643

Ettlie JE, Reza EM (1992) Organizational integration and process innovation. Acad Manage J 35(4):795–827

Evan WM (1966) The organization set: toward a theory of interorganizational relations. In: Thompson JD (ed) Approaches to organizational design. University of Pittsburgh Press, Pittsburgh

Evangelista R, Vezzani A (2010) The economic impact of technological and organizational innovations a firm-level analysis. Res Policy 39(10):1253–1263

Fleck J (1994) Learning by trying: the implementation of configurational technology. Res Policy 23(6):637–652

Hamel G (2006) The why, what, and how of management innovation. Harv Bus Rev 84(2):72–84

Herbst PG (1974) Socio-technical design: strategies in multidisciplinary research. Tavistock Publications, London

Laursen K, Salter A (2006) Open for innovation: the role of openness in explaining innovation performance among UK manufacturing firms. Strateg Manage J 27(2):131–150

Leonard-Barton D, Deschamps I (1988) Managerial influence in the implementation of new technology. Manage Sci 34(10):1252–1265

Luria D (1987) Technology, work organization, and competitiveness: automotive subsystem cost reduction, 1986–1992. Industrial Technology Institute, University of Michigan

Mol M, Birkinshaw J (2009) The sources of management innovation: when firms introduce new management practices. J Bus Res 62(12):1269–1280

Mundlak Y (1961) Aggregation over time in distributed lag models. Int Econ Rev 2(2):154–163

OECD (2005) Oslo manual: guidelines for collecting and interpreting innovation data. OECD, Paris

Pasmore W, Francis C, Haldeman J, Shani A (1982) Sociotechnical systems: a north American reflection on empirical studies of the seventies. Hum Relat 35(12):1179–1204

Roberts PW, Amit R (2003) The dynamics of innovative activity and competitive advantage: the case of Australian retail banking, 1981 to 1995. Organ Sci 14(2):107–122

Ruigrok W, Peck S, van der Linde C (1999) Strange bedfellows! Foreigners on top management teams and supervisory boards. Annual meeting of the strategic management society, Berlin, Germany

Schmidt T, Rammer C (2007) Non-technological and technological innovation: strange bedfellows? ZEW Discussion Papers 07-052

Scott WR (1992) Organizations: natural, rational and open systems. Prentice-Hall, Englewood Cliffs

Tether BS (2002) Who co-operates for innovation, and why: an empirical analysis. Res Policy 31(6):947–967

Trist EL (1978) On socio-technical systems. Sociotechnical systems: a sourcebook. University Associates, San Diego

Trist EL, Bamforth KW (1951) Some social and psychological consequences of the longwall method: an examination of psychological situation and defences of a work group in relation to the social structure and technological content of the work system. Hum Relat 4(3)

Trist E, Murray H, Trist B (1993) The social engagement of social science: a Tavistock anthology, vol. II: The socio-technical perspective. University of Pennsylvania Press, Philadelphia

Vaccaro IG, Jansen JJP, Van Den Bosch FAJ, Volberda HW (2012) Management innovation and leadership: the moderating role of organizational size. J Manage Stud 49(1):28–51

Voss CA (1988) Implementation: a key issue in manufacturing technology: the need for a field of study. Res Policy 17(2):55–63

Walker RM, Damanpour F, Devece CA (2011) Management innovation and organizational performance: the mediating effect of performance management. J Public Adm Res Theory 21(2):367

Wengel J, Lay G, Nylund A, Bager-Sjögren L, Stoneman P, Bellini N, Bonaccorsi A, Shapira P (2000) Analysis of empirical surveys on organisational innovation and lessons for future community innovation surveys. Scientific Follow-Up of the Community Innovation Survey (CIS) Project CIS 98(191):483

Womack JP, Jones DT, Roos D (1990) The machine that changed the world: the story of lean production. Rawson Associates, New York

Chapter 3
Environmental Performance: Interplay Between the Roles of Process Innovation Capability and Managerial Innovation Implementation

Deepa Aravind, Fariborz Damanpour, and Carlos Devece

Abstract Unlike a large portion of previous innovation research that has focused on performance consequences of the adoption of technological innovations, this study focuses on the benefits gained from the implementation of managerial innovations. We examine the mediating role of managerial innovation implementation on the influences of technological process innovation capability and pressure from external stakeholders on environmental performance. We develop hypotheses and test them using survey data from 192 ISO 14001 certified facilities in the US. The results suggest that extent of implementation partially mediate the effect of process innovation capability but fully mediate the effect of external pressure on environmental performance. We discuss the implications of these findings for research and practice on the implementation of managerial innovations.

3.1 Introduction

This study aims to address two research needs on the adoption and consequences of innovation in organizations. First, prior research agrees that innovation by firms in their products and processes is critical in maintaining a competitive advantage over

D. Aravind (✉)
College of Staten Island – School of Business, City University of New York, Staten Island, NY, USA
e-mail: deepa.aravind@csi.cuny.edu

F. Damanpour
Department of Management and Global Business, Rutgers University, Newark, NJ, USA
e-mail: damapour@business.routgers.edu

C. Devece
Department of Business Organization, Universitat Politècnica de València, Valencia, Spain
e-mail: cdevece@upvnet.upv.es

J.-L. Hervás-Oliver and M. Peris-Ortiz (eds.), *Management Innovation*, Springer Proceedings in Business and Economics, DOI 10.1007/978-3-319-03134-7_3,
© Springer International Publishing Switzerland 2014

rivals and is important for improving performance and generating growth not only for firms that develop them, but also for firms that adopt and use them (Damanpour and Aravind 2006). However, compared to the generation and adoption, the implementation of innovation has garnered much less research attention. Very often firms are unable to derive benefits out of an innovation not because of any inherent drawbacks in the innovation itself but because of implementation failure (Klein and Sorra 1996; Repenning 2002). Also, for an innovation to deliver improvements and actually affect organizational outcomes, its implementation should be effective through the acceptance and regular use of the innovation by organizational members and other constituents (Klein and Sorra 1996; Walker et al. 2011). Second, studies of managerial innovation, compared with technological product and process innovations, have been scarce despite apparent agreement among researchers that managerial innovations also play a crucial role in affecting firm competitiveness, growth, and performance (Damanpour and Aravind 2012; Edquist et al. 2001; Birkinshaw et al. 2008). Particularly, studies examining the effect of extent of implementation of managerial innovation on organizational outcomes are not many (Douglas and Judge 2001) and evidence regarding such effects is mixed. Some studies have found positive effects on performance dimensions such as financial performance and operating performance (Naveh et al. 2006) and others have not found such effects (Staw and Epstein 2000).

We aim to address these research needs by examining the extent of implementation of the ISO 14001, an environmental management system (EMS), on firm environmental performance. We define the extent of implementation of ISO 14001 as the degree to which the firm adheres to the requirements of the EMS and embeds the activities prescribed by the ISO 14001 standard in its daily routines (Aravind and Christmann 2011). Firms that implement to a lesser extent fail to continuously comply with the EMS's requirements or use the prescribed activities in their daily operations, while those that implement to a higher extent consistently comply with the requirements of the EMS and embed the prescribed activities into their daily routines.

We extend and expand the existing research by proposing a more nuanced model on the effects of external pressure on a firm for being environmentally responsible and the firm's technological process innovation capability on environmental performance and examine the mediating role of the extent of implementation of ISO 14001 EMS on these relationships. In particular, we address three research questions: (1) does the extent of implementation affect performance; (2) does the extent of implementation mediate the relationship between technological process innovation capability and performance; and (3) does the extent of implementation mediate the relationship between external pressure and performance? By focusing on these questions, our study aims at making two important contributions: (1) to add to the scant empirical research on performance implications of the implementation of managerial innovation; and (2) to develop a conceptual model that better highlights the interplay between technological process innovation and managerial innovation implementation on affecting performance.

The ISO 14001 environmental management system (EMS) is particularly suited for our study. While it is often difficult to measure the performance impacts of a managerial innovation (Damanpour and Aravind 2012), performance consequences of ISO 14001 EMS can be more easily measured as its implementation affects one specific aspect of organizational performance, namely, environmental performance. We use a questionnaire survey as our data source, supplemented with secondary data sources like the Dunn and Bradstreet and Quality Systems Update (QSU) databases. Our analysis is based on 192 survey responses obtained from facility managers in the US who are responsible for ISO 14001 in their facilities. We test our hypotheses using Structural Equation Modeling (SEM), which is a rigorous procedure for analyzing mediating effects (Bollen 1987; Cheung 2007). The results generally suggest that while quality of implementation partially mediates the effect of process innovation capability on environmental performance, it fully mediates the influence of external pressure.

3.2 Theory

3.2.1 Innovation Types

Innovation is generally defined as the development or use of new products, services, production processes, organizational structures, or administrative systems that are new to the adopting firm (Damanpour and Aravind 2012; Rogers 1995; Walker et al. 2011). Innovation researchers have distinguished between managerial and technological innovations. Managerial innovations are new organizational structures, administrative systems, management practices, processes, and techniques that could create value for the organization (Birkinshaw et al. 2008; Damanpour and Aravind 2012; Kimberly 1981). They are departures from traditional management principles, processes, and practices that alter 'the way the work of management is performed' and change 'how managers do what they do' (Hamel 2006, pp. 75–76). The primary purpose of managerial innovations is to improve the efficiency of internal operational and administrative processes. In this study, we focus on the implementation of managerial innovation. The implementation of innovation has been defined as "the process of gaining targeted organizational members' appropriate and committed use of an innovation" (Klein and Sorra 1996, p. 1055). Innovation implementation is preceded by innovation adoption, a decision that is typically made by senior managers on the assumption that the adopted innovation will be used by the members of the organization.

Technological innovations are directly related to the primary work activity of the organization and produce changes in its outputs and operating systems. Product and process innovations are conceived as two types of technological innovations. Product innovations are defined as new products or services introduced to meet an external user need, and process innovations are defined as new elements introduced into a firm's production or service operations to produce a product or render a service (Damanpour and Aravind 2006; Schilling 2008; Utterback 1994). While product innovations are "products or services that are new to the market", process innovations are "new ways of manufacturing existing or new products" (Damanpour and Aravind 2006, p. 41). The drivers of product innovations are primarily customer demand for new products and managers' desire to penetrate new markets, whereas the drivers of process innovations are mainly reduction in delivery lead time, lowering of operational costs, and increase in flexibility (Boer and During 2001; Schilling 2008). Technological process innovations change the way products are produced by introducing change in technology related to equipment and machinery, operational techniques, and production systems (Boer and During 2001; Damanpour and Aravind 2006; Utterback 1994). In this study, we examine joint effects of technological process and managerial innovations on performance.

3.2.2 ISO 14001 Environmental Management System

We investigate the ISO 14001 EMS, which, along with other similar management standards like the ISO 9000 Quality Management System, has been conceptualized as a managerial innovation (MI) in previous studies (Henriques and Sadorsky 2007; Naveh et al. 2006). The ISO 14001 EMS is a formal system intended to help organizations manage their environmental issues such as waste and emissions. It was established by the International Organization for Standardization in 1996 and is the most widely adopted certifiable EMS standard with more than 250,972 certified facilities in 155 countries as of December 2010 (http://www.iso.org/iso/iso-survey2010.pdf, accessed January 2012).

The implementation of ISO 14001 requires reviewing environmental issues regularly, ascertaining what needs to be done to minimize environmental impacts, providing continuous training for employees, conducting regular reviews and internal audits, and maintaining documentation that will help in codifying the facility's environmental knowledge. The standard specifies process requirements for the design of the EMS. Firms can obtain ISO 14001 certification of their EMS by passing an audit conducted by independent accredited third-party registrars and need to be recertified every 3 years. The implementation of EMS involves significant internal organizational changes involving changes in environmental policies, goals, strategies, and administrative procedures (Coglianese and Nash 2001).

3.2.3 Environmental Performance

Environmental performance is a critical dimension of organizational effectiveness that needs to be considered by firms (Judge and Douglas 1998). The expectation is that firms' adherence to the formal requirements of ISO 14001 will lead to changes

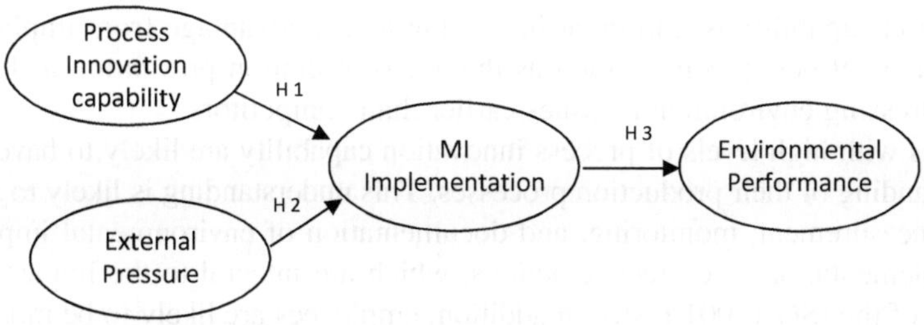

Fig. 3.1 Hypothesized model

in how environmental issues are managed that can result in improvements in their environmental performance. First, requirements of the standard such as the senior management commitment, establishment of environmental policies, involvement and training of employees and managers, and regular tracking of firms' environmental performance and progress towards the achievement of its environmental goals contribute to environmental performance improvements. Second, the implementation of the ISO 14001 EMS often promotes the adoption of additional environmental practices, such as recycling systems and the acquisition of clean technology (Gonzalez-Benito and Gonzalez-Benito 2008; Sroufe 2003). Third, ISO 14001's third-party audit system provides a monitoring mechanism that is intended to ensure that certified firms comply with the ISO 14001 requirements. It is intended to assess the extent to which firms comply with ISO requirements and to help spot opportunities for improvement (Jiang and Bansal 2003). Thus, if implemented well, ISO 14001 has the potential of improving a firm's environmental performance.

Figure 3.1 provides our hypothesized model. We propose that a firm's process innovation capability as well as external pressure to be environmentally responsible affect environmental performance, but through the extent of implementation of ISO 14001. Given the findings in extant literature on the performance implications of both process innovation (Augusto et al. 2011; Prajogo 2006) and external pressure (Lee 2009), we have included these variables as primary antecedents in our model. For instance, Augusto et al. (2011) found that process innovation was positively related to firm performance, and Prajogo (2006) reported that in manufacturing firms, process innovation had a stronger relationship with firm performance than product innovation. Regarding external pressure, Lee (2009) found that public corporations, as opposed to private ones, were exposed to greater external pressure, and that such pressure led to consistently better environmental performance.

3.2.4 Process Innovation Capability and MI Implementation

Process innovation capability is defined as a firm's ability to introduce innovations in its production processes. This capability is important in gaining the most out of the adoption of new practices. For instance, Christmann (2000) found that process innovation capability is important in gaining a cost advantage from implementing environmental best practices such as the use of pollution prevention technologies and addressing environmental issues earlier than competitors.

Firms with high levels of process innovation capability are likely to have a better understanding of their production processes. This understanding is likely to aid them in the measurement, monitoring, and documentation of environmental impacts and the implementation of corrective actions, which are integral to the implementation process of the ISO 14001 EMS. In addition, employees are likely to be more skilled and involved, which fosters a greater level of implementation of an EMS (Whitelaw 2004). Therefore, firms with higher process innovation capability are likely to implement the EMS to a higher extent. Such firms are also more likely to gain competitive advantage out of adopting ISO 14001 as opposed to those firms that implement it to a lesser extent. The above discussion leads to the following hypothesis:

Hypothesis 1: Process innovation capability is positively related to the extent of implementation of the ISO 14001 EMS standard.

3.2.5 External Pressure and MI Implementation

External pressure plays a critical role in shaping firm behavior. For instance, institutional theory suggests that firms face significant pressures from external constituents such as regulatory agencies, industrial associations, and other stakeholders to adopt practices and structures (Meyer and Rowan 1977; Scott 1987; Zucker 1987). If firms do not conform to these institutional demands, they risk losing their legitimacy, social support, access to resources and other benefits that are associated with conformance (Meyer and Rowan 1977; Scott 1987; Zucker 1987).

In the case of ISO 14001, firms often face pressures from various sources to be environmentally responsible. Such sources include regulatory agencies, neighboring community, customers, shareholders, non-governmental organizations, and so on (Henriques and Sadorsky 2007). Nearly all industrialized countries now regulate industrial wastes and emissions and firms must undertake pollution control activities in order to remain within the law (Henriques and Sadorsky 2007). Thus, in many countries firms face significant pressures to conform to environmental regulations. Firms also face pressures from neighboring community and environmental organizations. These pressures can also be quite significant as these groups often protest against environmental emissions that reduce their quality of life. Failure to conform to these pressures could result in negative consequences to the firm.

Studies have suggested that external pressures can induce a firm to work harder at trying to reduce their environmental effects thereby resulting in better quality of EMS implementation. For instance, Sharma and Henriques (2005) found that even when corporate headquarters did not require it, facilities often exceeded governmental regulations pertaining to the environment due to community pressures. Firms that faced higher external pressures are more likely to respond to such pressures by modifying their behaviors. Therefore, we hypothesize:

Hypothesis 2: External pressure is positively related to the extent of implementation of the ISO 14001 EMS standard.

3.2.6 MI Implementation and Environmental Performance

For an innovation to deliver improvements and actually affect organizational performance, it is necessary for it to be implemented, through the acceptance and regular use of the innovation by employees and other constituents (Walker et al. 2011). Existing evidence on the relationship between implementation of managerial innovation and organizational performance generally points to a positive effect between the two constructs (Camison and Lopez 2010; Han et al. 1998; Ho 2010; Mol and Birkinshaw 2009). For example, in a study of best managerial practices, Camison and Lopez (2010) found that implementation of managerial innovations is positively related to both economic performance and stakeholder satisfaction. Ho (2010), in a study of implementation of strategy, structure, systems, and culture, also found a positive relationship between managerial innovation and both financial and market performance. Similarly, Mol and Birkinshaw (2009) found that the implementation of new managerial practices is positively related to firm financial performance.

With regard to the ISO 14001 EMS, if firms implement the system to a high extent, this can lead to changes in their management of environmental issues, which can result in improvements in their environmental performance. For example, senior management commitment, the establishment of environmental policies, involvement and training of employees and managers, and documentation of environmental practices can contribute to environmental performance improvements. Such practices help integrate environmental concerns into daily activities, raise management and employee awareness, and add more rigor to environmental programs (Jiang and Bansal 2003).

Extant literature that examines the relationship between the extent of implementation of managerial innovation and environmental performance is rare, though more recently a few studies have examined this link. Yin and Schmeidler (2009) found that managers in certified facilities with low quality of the ISO 14001 implementation believe that ISO 14001 does not result in environmental performance benefits. In a study based primarily on survey data on the implementation of the ISO 9000 standard, Naveh et al. (2006) found that the extent of implementation of ISO 9000 is positively related to both operational and financial performance improvements. Using an external measure for environmental performance, Aravind and Christmann (2011) found that "ISO 14001 certified firms that implemented to a higher extent" have better environmental performance than their non-certified counterparts. Based on the above evidence, we suggest that:

Hypothesis 3: The extent of implementation of the ISO 14001 EMS standard is positively related to environmental performance.

3.3 Methods

3.3.1 Sample and Data

We tested our hypotheses using data from a mail questionnaire survey of ISO 14001 certified facilities in the United States (Aravind and Christmann 2011). We used this method because data on the implementation of ISO 14001 in facilities cannot be obtained from public sources. We obtained a list of 5,284 ISO certified facilities from QSU Publishing Company's ISO 14001 Worldwide Certified Company Directory (QSU 2006), the most comprehensive database of certified facilities in the United States. To ensure that we would be able to perform adequate follow-up to the survey resulting in a good response rate, we restricted our mailing sample to 600 randomly selected facilities from the QSU directory.

The target respondent in our survey was the individual who was responsible for ISO 14001 EMS at the facility and most knowledgeable about the implementation of the system. This approach is the principal methodological solution to using single respondents (Campbell 1955; John and Reeves 1982), as the respondent can validly assess the construct (Crampton and Wagner 1994). The survey questionnaire was developed based on existing literature and incorporated feedback from managers. We based the survey administration on the tailored design method which has been shown to improve response rates to mail survey questionnaires (Dillman 2000).

Two weeks after our first mailing, we sent a reminder letter and followed this with two more mailings. Of the 600 mailed surveys, 13 were undeliverable due to incorrect addresses, and of the remaining 587 surveys, 199 were returned completed, yielding an overall response rate of 33.9 %. Due to incomplete data only 192 of these responses were usable for testing our model. The median size of our respondent facilities was 200 employees with the number of employees ranging from 6 to 2,700. The facilities were on average 5.2 years ISO 14001 certified with a minimum of 1 year and a maximum of 11 years.

We performed several tests including wave analysis (Fowler 1993) to ensure that the respondents were representative of our mailing sample. Also, we tried as much as possible to reduce common method bias or at least estimate its extent. First, we guaranteed anonymity to the respondents and reduced evaluation apprehension by assuring respondents that there are not right or wrong answers. Second, we made every effort to improve our questionnaire items by avoiding vague concepts and keeping questions simple and precise. Third, we conducted Harman's single factor test (Podsakoff et al. 2003) for the extent of common method variance. Multiple factors with eigenvalues greater than 1.0 emerged from these analyses that accounted for 66 % of the total variance. No single factor accounted for a majority of the variance in the data. Together, these suggest that common method bias could not account for all the relationships among scale items in our data set and is not a serious problem in this study.

Table 3.1 Means, standard deviation and correlations

			Correlations			
Variables	Mean	SD	PIC	MII	EP	ExPr
PIC (Process innovation capability)	4.92	1.20	(.87)			
MII (Extent of MI implementation)	5.32	1.08	.28**	(.81)		
EP (Environmental performance)	5.38	.85	.382*	.56**	(.88)	
ExPr (External pressure)	3.55	1.19	.06	.26**	.28**	(.72)

Note: Reliability coefficients (Cronbach's alpha) for the scales are in parenthesis
n=192; *p<.05; **p<.01 (two tailed)

3.3.2 *Measures*

Appendix 1 lists the questions used to construct our measures. Reliability coefficients for all multi-item scales are provided in Table 3.1.

Process innovation capability (PIC) was measured by four items and is based on Christmann (2000). We used the stem question "relative to your major competitors, your facility focuses on" followed by items including "being a leader in process innovation" and "capital investment in new equipment and machinery".

Our measure of extent of implementation of ISO 14001 (MII) was based on four items used in previous studies (Christmann and Taylor 2006; Naveh and Marcus 2004, 2005). We used the stem question "this question pertains to the implementation and perceptions of the ISO 14001 EMS at your facility. To what extent" followed by items including "are the documents created for the purpose of ISO 14001 used in daily practice?" and "is the system regularly ignored?". A low score indicates low extent of MI implementation and a high score indicates a high extent.

We measured environmental performance (EP) using six survey items based on Judge and Douglas (1998) and Chan (2005). The stem question was "how does your facility compare with other facilities in your industry along the following dimensions?", followed by items including "limiting environmental impact beyond complying with regulations" and "preventing environmental accidents".

Our measure for external pressure (ExPr) used four survey items: the stem question was "how strong were the following forces in pressuring your facility to be environmentally responsible" and was followed by regulatory agencies, neighbours of your facility, environmental NGOs and immediate suppliers of your facility. Table 3.1 provides descriptive statistics and correlations for all the variables used in the analysis.

3.3.3 *Analysis*

We tested our hypotheses using Structural Equation Modeling, employing the EQS 6.1 software to estimate the model (Bentler and Wu 1995). We used the maximum-likelihood method combined with the method of robust standard estimators to avoid restrictions on the normality of the data (Satorra and Bentler 2001). Our hypothesized model shown in Fig. 3.1 consists of two exogenous variables (process innovation capability and external pressure) and two endogenous variables (extent of MI implementation and environmental performance). The overall fit of the models was evaluated by a combination of absolute, incremental, and parsimonious fit indexes recommended by Hair et al. (1992) and Jöreskog and Sorbom (1993).

3.4 Results

The relationships established in the hypotheses were jointly assessed by the structural model (Model 1) and are shown in Fig. 3.2. Applying the maximum-likelihood method, all the indices indicate that the hypothesized model has adequate fit to the data (NNFI = 0.90; CFI = 0.92, IFI = 0.92; NC = 1.72; RMSEA = 0.06). Figure 3.2 also contains the parameter estimates for the main predictors, significance levels, and proportions of explained variance (R^2).

The results support all the hypotheses. Hypotheses 1 and 2 were confirmed as the path coefficient of PIC (.38) and ExPr (.33) on MII are positive and significant ($p < .01$), explaining jointly 25 % of the MII variance. Hypothesis 3 was also supported as the effect of MII on EP is significant ($p < .01$) and the R^2 of EP is 0.36.

3.5 Conclusions

Strategy scholars suggest that innovating across the organization would enable an organization to renew its ability to build, reconfigure, and integrate internal and external competencies to cope with environmental change and remain effective over time (Eisenhardt and Martin 2000; Van den Bosch et al. 1999). Also, the

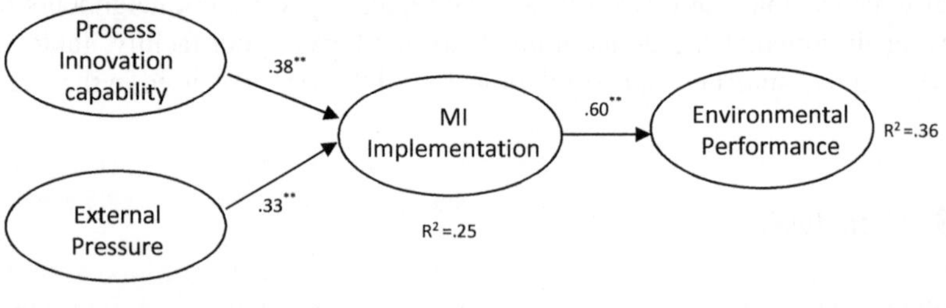

Fig. 3.2 Path coefficients for the hypothesized model $\chi^2 = 256.7$; d.f. = 149; NNFI = 0.90; CFI = 0.92, IFI = 0.92; NC = 1.72; RMSEA = 0.06 * $p < .05$; ** $p < .01$

resource-based view suggests that the use of the organization's internal resources, including technological and managerial knowledge resources, lead to performance consequences (Barney 1991; Damanpour et al. 2009). The importance of both technological and managerial innovations for firm performance and economic growth have also been noted in the economic literature (Lazonick 1991; Nelson 1996; Sanidas 2005).

In this study, we examined ISO 14001 EMS as a managerial innovation and investigated whether the extent of its implementation mediates the relationship between technological process innovation capability and organizational performance. Additionally, we examined how the influence of external pressures on organizational performance is affected by the extent of implementation of ISO 14001. We used Structural Equation Modeling to test our hypotheses on a sample of ISO 14001 EMS certified facilities in the US and found that the extent of managerial innovation implementation (1) partially mediates the relationship between technological process innovation capability and environmental performance, and (2) fully mediate the relationship between external pressure and environmental performance.

These findings make several contributions to the literature. First, we address the imbalance in the innovation literature where technological product innovation has received much of the scholarly attention to the detriment of technological process and managerial innovations (Keupp et al. 2012). Second, we focus on the role of implementation of managerial innovation that has not received much research attention. In the environmental literature, studies of environmental management systems have not considered implementation issues much and have almost exclusively considered adoption or the act of certification (Darnall 2001, 2003; King et al. 2005). Prior innovation research suggests that if an innovation is not implemented, it will not affect firm processes and behavior and will not influence performance outcomes. Thus, ineffective implementation could be a major cause for the inability of firms to improve their performance through innovation adoption (e.g., Klein and Sorra 1996). Accordingly, we found that the extent of managerial innovation implementation does indeed have a significant impact on environmental performance. In this vein, our findings underscore the need for practicing managers to be cognizant of implementation issues and allocate necessary resources to ensure effective implementation of managerial innovations to attain desired performance outcomes.

In summary, this study suggests that a better understanding of performance consequences of innovation adoption requires developing and testing more complex models that account for the interplay between managerial innovation implementation with process innovation capability and external pressure for environmental responsibility. Future research could contribute by including additional variables to our model, testing other types of managerial innovation, utilizing different analytical methods, and developing entirely new models to examine the performance consequences of managerial innovation in organizations.

3.6 Appendix 1: Measures

Construct	Indicator
Environmental performance	*Survey items: (rated on 7-point Likert scale)* How does your facility compare to other facilities in your industry along the following dimensions? *(Much below average..Average..Much above average)* 1. Complying with environmental regulations 2. Limiting environmental impact beyond complying with regulations 3. Preventing environmental accidents 4. Lessening the impact of environmental accidents 5. Educating employees about the environment 6. Environmental performance
Extent of MI implementation	*Survey items: (rated on 7-point Likert scale)* This question pertains to the implementation and perceptions of the ISO 14001 EMS at your facility. To what extent *(Not at all..To a large extent)*: 1. Are the documents created for the purpose of ISO 14001 used in daily practice? 2. Has the ISO 14001 system become part of your regular routine? 3. Are preparations for external audits made at the last minute? (reverse-scored) 4. Is the system regularly ignored? (reverse-scored) (**Correlated** this measure with average emissions data from US TRI database – obtained a significant negative correlation)
Process innovation capability	*Survey items: (rated on 7-point Likert scale)* Relative to your major competitors, your facility focuses on *(Strongly disagree..Strongly agree)*: 1. Being the first in the industry to try new methods and technologies 2. Using the latest technology in production 3. Capital investment in new equipment and machinery 4. Being a leader in process innovation
External pressures	*Survey items: (rated on 7-point Likert scale)* How strong are the following forces in pressuring your facility to be environmentally responsible? *(No pressure..Moderate pressure..Intense pressure)* 1. Immediate suppliers of your facility 2. Regulatory agencies 3. Environmental non-governmental organizations (NGOs) 4. Neighbors of your facility

References

Aravind D, Christmann P (2011) Decoupling of standard implementation from certification: does quality of ISO 14001 implementation affect facilities' environmental performance? Bus Ethics Q 21(1):75–104

Augusto M, Lisboa J, Yasin M (2011) The impact of innovation on the relationship between manufacturing flexibility and performance: a structural modelling approach. Int J Bus Res 11(4): 65–72

Barney J (1991) Firm resources and sustained competitive advantage. J Manage 17(1):99–120

Bentler PM, Wu EJC (1995) EQS for Windows: user's guide. Multivariate Software, Encino

Birkinshaw J, Hamel G, Mol MJ (2008) Management innovation. Acad Manage Rev 33(4): 825–845

Boer H, During W (2001) Innovation, what innovation? A comparison between product, process and organisational innovation. Int J Technol Manage 22(1–3):83–107

Bollen KA (1987) Total, direct, and indirect effects in structural equation models. In: Clogg CC (ed) Sociological methodology. American Sociological Association, Washington, DC, pp 37–69

Camison C, Lopez AV (2010) An examination of the relationship between manufacturing flexibility and firm performance: the mediating role of innovation. Int J Oper Prod Manage 30(8): 853–878

Campbell DT (1955) The informant in quantitative research. Am J Sociol 60(4):339–342

Chan RYK (2005) Does the natural-resource-based view of the firm apply in an emerging economy? A survey of foreign investment enterprises in China. J Manage Stud 42:625–666

Cheung MWL (2007) Comparison of approaches to constructing confidence intervals for mediating effects using structural equation models. Struct Eq Model 14:227–246

Christmann P (2000) Effects of best practices of environmental management on cost advantage: the role of complementary assets. Acad Manage J 43(4):663–680

Christmann P, Taylor G (2006) Firm self-regulation through international certifiable standards: determinants of symbolic versus substantive implementation. J Int Bus Stud 37(6):863–878

Coglianese C, Nash J (2001) Environmental management systems and the new policy agenda. In: Coglianese C, Nash J (eds) Regulating from the inside can environmental management systems achieve policy goals? Resources for the Future, Washington, DC, pp 1–26

Crampton SM, Wagner MA (1994) Percept-percept inflation in microorganizational research. J Appl Psychol 79(1):67–76

Damanpour F, Aravind D (2006) Product and process innovations: a review of organizational and environmental determinants. In: Hage J, Meeus M (eds) Innovation, science, and institutional change. Oxford University Press, Oxford, pp 38–66

Damanpour F, Aravind D (2012) Managerial innovation in organizations: conceptions, processes and antecedents. Manage Organ Rev 8(2):423–454

Damanpour F, Walker RM, Avellaneda CN (2009) Combinative effects of innovation types on organizational performance: a longitudinal study of public services. J Manage Stud 46(4): 650–675

Darnall N (2001) Adopting ISO 14001: why some firms mandate certification while others encourage it. Paper presented at the 23rd annual research conference for the association for public policy analysis and management fall conference, Washington, DC

Darnall N (2003).Why U.S. firms certify to ISO 14001: an institutional and resource-based view. Paper presented at the best paper proceedings of the academy of management conference, Seattle, Washington, DC

Dillman DA (2000) Mail and internet surveys: the tailored design method. Wiley, New York

Douglas TJ, Judge WQ (2001) Total quality management implementation and competitive advantage: the role of structural control and exploration. Acad Manage J 44(1):158–169

Edquist C, Hommen CL, McKelvey M (2001) Innovation and employment: process versus product innovation. Edward Elgar, Cheltenham

Eisenhardt K, Martin JA (2000) Dynamic capabilities: what are they? Strateg Manage J 21(10–11): 1105–1121

Fowler FJS Jr (1993) Survey research methods. Sage, Newbury Park

Gonzalez-Benito J, Gonzalez-Benito O (2008) Operations management practices linked to the adoption of ISO 14001: an empirical analysis of Spanish manufacturers. Int J Prod Econ 113(1):60–73

Hair JF, Anderson RE, Tatham RL, Black WC (1992) Multivariate data analysis. Macmillan, New York

Hamel G (2006) The why, what and how of management innovation. Harv Bus Rev 84(2):72–84

Han JK, Kim N, Srivastava RK (1998) Market orientation and organizational performance: is innovation the missing link? J Mar Sci 62(4):30–45

Henriques I, Sadorsky P (2007) Environmental technical and administrative innovations in the Canadian manufacturing industry. Bus Strateg Environ 16(2):119–132

Ho L (2010) Meditation, learning, organizational innovation, and performance. Ind Manage Data Syst 111(1):113–131

Jiang RJ, Bansal P (2003) Seeing the need for ISO 14001. J Manage Stud 40(4):1047–1067

John G, Reeves T (1982) The reliability and validity of key informant data from dyadic relationships in marketing channels. J Mark Res 19:517–525

Joreskog KG, Sorbom D (1993) LISREL 8 user's reference guide. Scientific Software International, Chicago

Judge WQ, Douglas TJ (1998) Performance implications of incorporating natural environmental issues into the planning process: an empirical assessment. J Manage Stud 35(2):241–262

Keupp MM, Palmié M, Gassmann O (2012) The strategic management of innovation: a systematic review and paths for future research. Int J Manage Rev 14(4):367–390

Kimberly JR (1981) Managerial innovation. In: Nystrom PC, Starbuck WH (eds) Handbook of organizational design. Oxford University Press, New York, p 104

King AA, Lenox MJ, Terlaak A (2005) The strategic use of decentralized institutions: exploring certification with the ISO 14001 management standard. Acad Manage J 48(6):1091–1106

Klein KJ, Sorra JS (1996) The challenge of innovation implementation. Acad Manage Rev 21(4): 1055–1080

Lazonick W (1991) Business organizations and the myth of the market economy. Cambridge University Press, New York

Lee MP (2009) Does ownership form matter for corporate social responsibility? A longitudinal comparison of environmental performance between public, private, and joint-venture firms. Bus Soc Rev 114(4):435–456

Meyer JW, Rowan B (1977) Institutionalized organization: formal structure as myth and ceremony. Am J Sociol 83(2):340–363

Mol MJ, Birkinshaw J (2009) The sources of management innovation: when firms introduce new management practices. J Bus Res 62(12):1269–1280

Naveh E, Marcus AA (2004) When does the ISO 9000 quality assurance standard lead to performance improvement? Assimilation and going beyond. IEEE Trans Eng Manage 51(3): 352–363

Naveh E, Marcus A (2005) Achieving competitive advantage through implementing a replicable management standard: installing and using ISO 9000. J Oper Manage 24(1):1–26

Naveh E, Meilich O, Marcus AA (2006) The effects of administrative innovation implementation on performance: an organizational learning approach. Strateg Organ 4(3):275–302

Nelson RR (1996) The evolution of comparative or competitive advantage: a preliminary report on a study. Ind Corp Change 5(2):597–617

Podsakoff PM, MacKenzie SB, Lee JY, Podsakoff NP (2003) Common method biases in behavioral research: a critical review of the literature and recommended remedies. J Appl Psychol 88(5):879

Prajogo DI (2006) The relationship between innovation and business performance—a comparative study between manufacturing and service firms. Knowl Process Manage 13(3):218–225

QSU (2006) ISO 14001 Worldwide Certified Company Directory
Repenning NP (2002) A simulation-based approach to understanding the dynamics of innovation implementation. Organ Sci 13(2):109–127
Rogers EM (1995) Diffusion of innovations. Free Press, New York
Sanidas E (2005) Organizational innovations and economic growth: organosis and growth of firms, sectors, and countries. Edward Elgar, Cheltenham
Satorra A, Bentler PM (2001) A scaled difference chi-square test statistic for moment structure analysis. Psychometrika 66(4):507–514
Schilling MA (2008) Strategic management of technological innovation. McGraw-Hill Irwin, New York
Scott RW (1987) The adolescence of institutional theory. Adm Sci Q 32(4):493–511
Sharma S, Henriques I (2005) Stakeholder influences on sustainability practices in the Canadian forest products industries. Strateg Manage J 26(2):159–180
Sroufe RP (2003) Effects of environmental management systems on environmental management practices and operation. Prod Oper Manag 12(3):416–431
Staw BM, Epstein LD (2000) What bandwagons bring: effects of popular management techniques on corporate performance, reputation, and CEO pay. Adm Sci Q 45(3):523–556
Utterback JM (1994) Mastering the dynamics of innovation. Harvard Business Press, Cambridge, MA
Van den Bosch FAJ, Volberda HW, De Boer M (1999) Coevolution of firm absorptive capacity and knowledge environment: organizational forms and combinative capabilities. Organ Sci 10(5):551–568
Walker RM, Damanpour F, Devece CA (2011) Management innovation and organizational performance: the mediating effect of performance management. J Public Adm Res Theory 21(2): 367–386
Whitelaw K (2004) ISO 14001 environmental systems handbook, 2nd edn. Butterworth-Heinemann, Oxford
Yin H, Schmeidler P (2009) Why do standardized ISO 14001 environmental management systems lead to heterogeneous environmental outcomes? Bus Strateg Environ 18(4):469–486
Zucker LG (1987) Institutional theories of organizations. Annu Rev Sociol 13:443–464

Chapter 4
Understanding Organizational Innovation from Its Practice

Maria Larraza Malkorra

Abstract Organizational innovation is studied as an important source of competitive advantage both for firms and for territories. This relevance is related to the widening of the innovation concept, which is no longer limited to technology. However, the concept of organizational innovation is still considered ambiguous, and even the Oslo Manual recognizes that its referential definition is still at an exploratory level. The analysis of organizational innovation has been made further methodologically challenging by the fact that innovation processes are no longer understood as linear and predictive, but complex and variable. It follows that new studies and methods are required if we are to acquire a deeper knowledge of organizational innovation practices and their consequences for competitiveness. Consequently, this research has aimed to achieve a thorough understanding of how an organizational innovation process is developed and understood in practice; and to generate new theoretical insights about such processes for guiding future research. Grounded theory has been used as a suitable methodology for this inductive, longitudinal, field-based case study research. Already, preliminary results have helped us gain new theoretical insights about the suitability of the Oslo Manual's definition in practice and about the value of the application of an innovation generation and adoption process perspective to the study of organizational innovation. Work is still in-progress to consolidate first results, to guarantee their confirmability and to facilitate their transfer.

M. Larraza Malkorra (✉)
Faculty of Social and Human Sciences, University of Deusto, San Sebastian, Spain
e-mail: maria.larraza@deusto.es

J.-L. Hervás-Oliver and M. Peris-Ortiz (eds.), *Management Innovation*,
Springer Proceedings in Business and Economics, DOI 10.1007/978-3-319-03134-7_4,
© Springer International Publishing Switzerland 2014

4.1 Introduction

Even though over past decades research on innovation has mainly focused on product and process innovation, it is now commonly acknowledged that in fact innovation has to do with much more than just technology and R&D. In fact, in reality, other types of non-technological innovation have become increasingly relevant for competitiveness. These include organizational innovation, which is considered an important source of competitive advantage for firms, for an enterprise's knowledge development, and as a means of enabling other forms of innovation (Damanpour et al. 1989; Greenan 2003; Armbruster et al. 2007; Som et al. 2012).

In response to the broadening of the innovation concept, the Oslo Manual (OECD/Eurostat 2005) in its third edition enlarged its definition of innovation. Whereas in previous editions only technological innovations were recognised, now, its typology of innovations included that of "organizational innovation" taking up Schumpeter's original broad vision.

At the same time, there has been a paradigmatic shift in how the process of innovation generation within organizations is understood. Instead of the traditionally studied linear, sequential and predictable innovation processes, business practices have been found to be following complex models that include multiple feedback loops. This change in our understanding of innovation processes has brought with it an awareness that not all innovations need to be radical, but, rather, that they may involve incremental social and organizational changes, in addition to any necessary technological advances (Kline and Rosenberg 1986; Dodgson 2000; Rothwell 2003; Tidd and Bessant 2009; quoted in Som et al. 2012).

The innovation process is thus complex and variable, which makes analysis of organizational innovation methodologically challenging (Kirner et al. 2008). Moreover, the organizational innovation concept is still considered ambiguous (Lam 2005) and even the Oslo Manual (OECD/Eurostat 2005) recognizes that its definition is still at an exploratory level and recommends that further research should be carried out in order to deepen knowledge about its practice.

Consequently, with only a restricted understanding about organizational innovation, its diffusion and potential to transfer, new studies and approaches are required if we are to acquire deeper insights into its practice, and to be able to analyse consequences for business competitiveness.

The purposes of this research have been twofold: firstly, to deeply understand how an organizational innovation process is developed and understood in practice; and, secondly, to generate new theoretical insights about organizational innovation, which can guide further research on the topic.

4.2 Organizational Innovation

4.2.1 *The Framework for Studying Organizational Innovation*

The relevance of organizational innovation for innovation theory and practice was officially recognised in 2005, when the latest edition of the Oslo Manual of the European Commission and the OECD (which sets out the methodological basis for major innovation studies) for the first time included it among its innovation typologies and proposed for it a standard definition. However, the definition was recognised as exploratory, and it was understood that further and thorough about it was needed.

In fact, the concept relates two key business areas: organizations and innovation. The existing literature about each of them is voluminous and diverse, and shows that the relationship between organizations and innovation is complex, dynamic and multi-level. It was Lam (2005) who recognized this complexity and took on the challenge of studying it by considering three different, but related, aspects: the relationship between organizational structural forms and innovativeness; innovation as a process of organizational learning and knowledge creation; and organizational capacity for change and adaptation. Even though her study identified important overlaps and interconnections between these different aspects, it is clear that the strands of research on them have remained separate, with the consequence that there is now a lack of consensus about what would constitute an adequate conceptual framework for understanding organizational innovation.

Lam (2005) linked this omission to the ambiguity of the organizational innovation concept. In fact, research on the three aspects mentioned above has used, confusingly, the organizational innovation concept to refer to different phenomena, including: any innovation happening in an organizational context; the sum of innovations being developed by an organization; and innovation in organizational methods.

There is still plenty of work to do to understand how these different research strands fit together, but the focus of this research has been on Lam's third aspect, "organizational capacity for change and adaptation", which coincides with the Oslo Manual's definition: "the implementation of a new organizational method in the firm's business practices, workplace organization or external relations" (OECD/Eurostat 2005).

Significantly, in addition to proposing a core definition for organizational innovation, the Oslo Manual (OECD/Eurostat 2005) also added two key characteristics about this type of innovation to be considered: the novelty of the organizational method implemented, and the strategic reasons for its deployment. The occurrence, or not, of these two characteristics help to distinguish organizational innovation from mere organizational change. Thus, for example, for an organizational change to be considered an organizational innovation, it must be completely new to the organization. Furthermore, the mere statement of certain management strategies in a document cannot be taken as evidence of organizational innovation; its actual implementation in a firm's activities is a basic requirement. More recent studies have introduced a new defining criteria, specifying that strategic motivation is needed if a change is to be considered an innovation, one that results in a considerable improvement of competitive advantage and economic performance (Som et al. 2012). However, this new defining criteria has introduced more confusion because the organizational and management literatures also have their own definitions relating to strategic motivation for organizational change processes (Van de Ven 1992; Poole 2004).

Summarizing, this research has been looking at organizational innovation within a framework set by the innovation studies' literature, focussing on innovations in organizational methods that go beyond mere organizational change, and which are strategically oriented towards broad competitiveness improvement.

4.2.2 Deepening Understanding of the Process of Organizational Innovation

Following recognition of the increasing relevance of organizational innovation, many scholars have initiated empirical research to identify its effects and benefits (Armbruster et al. 2007; Kirner et al. 2008; Arraut 2009; Som et al. 2012). Recently, the benefits of organizational innovation have been classified along three dimensions (Som et al. 2012):

- Organizational innovation as a distinct form of innovation, which could directly result in substantial improvements in organizational performance.
- Organizational innovation as an enabler for other types of innovation, such as product, process or marketing innovation.
- Organizational innovation as a prerequisite for knowledge accumulation, enabling firms to increase the ability of its members to acquire, create and make the best use of competences, skills and knowledge.

Even though these three dimensions are extremely relevant for current firms' competitiveness, it is generally acknowledged that the empirical detailing of consequences or antecedents is not possible because of the fact that most organizational concepts confusingly address different aspects of business performance at the same time (Som et al. 2012).

In fact, the empirical basis for measuring organizational innovation is weak and scattered due to a lack of reliable measurement scales and because of the intangibility of the goals of this type of innovation (Armbruster et al. 2007). Furthermore, the measurement of the effects of organizational innovation is considered methodologically challenging for reasons connected to long life cycles, problems of aggregation, differences in the extent of implementation, and because of the multidimensional relationship between organizational innovation and its outcomes (Kirner et al. 2008).

Table 4.1 Models of the process of organizational change

Name	Change unit	Change mode	Description
Life cycle	Single unit	Planned	The change process of an organization conceived as a progression thorough a necessary sequence of stages that are prescribed by an institutional, natural or logical programme
Intentional	Single unit	Built	The change process conceived as a cycle of: vision formulation, implementation, evaluation, and modification based on what the organization has learnt
Dialectical	Multiple units	Built	The change process conceived as an evolution derived from constant confrontation and conflict
Evolutionary	Multiple units	Planned	The change process conceived as a repetitive sequence of variation, selection and retention of events, carried out by entities in a defined environment, whereby competition for scarce resources generates evolution

Source: Own elaboration based on Van de Ven (1992) and Poole (2004)

Management theory is helpful at this point, since its literature about organizational change includes theories of process and methods, which could be used to deepen understanding of organizational innovation. Instead of understanding change using a variation perspective (i.e. explaining change by relationships between dependent and independent variables, as studied by people using linear statistical models), Van de Ven (1992) proposed the use of a process perspective through which change is studied as a sequence of events, the researcher collecting narratives and carrying out longitudinal research. This approach requires qualitative research designs rather than quantitative ones.

Using the process perspective, Van de Ven (1992) classified four different models of organizational change utilising two variables: "change units" (i.e. either multiple units interacting and changing, or a single change unit), and a "change mode" (i.e. whether change progresses following a planned or unplanned built sequence). The resulting models, later developed further by Poole (2004), are described in Table 4.1.

The main value of this classification for organizational innovation research derives not from the four models themselves, which as they stand are inadequate for the study of organizational innovation since there are important elements that are not included, such as: degree of novelty, strategic motivation, and the possible applications of organizational innovation (namely business practices, workplace organization, and external relations). Rather, what is useful for carrying out organizational innovation process research is the idea of "innovation units" (e.g. organization wide or functionally limited units; and/or relating to only business practices, or also including external relations), and the idea of "innovation mode".

In the latter respect, there exists an extensive innovation literature about different modes of innovation processes, of which particularly relevant for this research has been that of Damanpour and Wischnevsky (2006), to the extent that it has similarities with the "built" or "planned" change mode variable used by Van de Ven (1992), and also by Poole (2004).

Damanpour and Wischnevsky's (2006) research was based on the perception that the existing literature about innovation in organizations did not usually distinguish between generation and adoption processes, referring to both equally as "innovation processes" (Daft 1982; Van de Ven 1986; Kanter 1988; Roberts 1988, 1995; quoted in Damanpour and Wischnevsky 2006). Consequently, Damanpour and Wischnevsky proposed distinguishing between the two types of innovation processes in the following terms.

The intention of the generation of innovations is that they contribute to an organization's effectiveness and competitiveness by creating new opportunities, or that they make use of an existing opportunity in novel ways (Drucker 1985). In contrast, the adoption of innovations refers to those developed elsewhere, such that when secondary organizations adopt them, they are assimilating products, services, or technologies new to the adopting organizations (Angle and Van de Ven 2000).

Damanpour and Wischnevsky (2006) concluded in their research that the processes involved in either generating or adopting innovation differed considerably. Summarizing, they consider, on the one hand, that the generation process covers all efforts and activities aimed at creating new ideas, making them work, and even to transferring them to other organizations. Generation is therefore a creative process in which new and existing ideas are combined in a novel way to produce an invention or a configuration that was previously unknown. The adoption process, on the other hand, is defined by two sub-processes: "initiation", including activities spanning the recognition of a need, to becoming aware of a possible innovation, to adapting this innovation to addressing the recognized need, to deciding to adopt it, and to planning its successful adoption; and "implementation", whereby the employment of the innovation is initiated and then continued in use until it becomes a routine part of the organization.

Given the above, Damanpour and Wischnevsky (2006) proposed that future research on innovation processes should distinguish between generation and adoption processes.

However, their empirical research was limited to cases of technological innovation, leaving a need to cover the distinction between generation and adoption processes in respect of organizational innovation. The authors considered that a development of innovation theory that takes into account generation and adoption differences could help clarify the problem of inconsistent research results about broad innovation issues (Damanpour and Wischnevsky 2006).

4.3 Methodology

Applying a process perspective to the study of organizational innovation requires an appropriate research design that properly addresses the methodological challenges posed by this phenomenon (Armbruster et al. 2007; Kirner et al. 2008). In this research the design proposed is based on an inductive, longitudinal, field-based case study, that is well suited to grounded theory (Eisenhardt 1989; Glaser and Strauss 1967).

In fact, it focuses on the reality and evolution of the field of study without following any prior hypotheses, and it seeks results in the collection and systematic analysis of field data. This approach is considered particularly useful for examining feedback processes that drive change dynamics over time (Tripsas 2009), such as those that may occur in organizational innovation processes. Moreover, grounded theory has been used for research in similar studies of strategy, organizational change and innovation processes (Van de Ven et al. 1989; Van de Ven and Huber 1990; Van de Ven 1992; Poole 2004).

4.3.1 Grounded Theory

Grounded theory has been designed as a methodology appropriate for developing theories based on data collected directly from the field. The theory obtained evolves during the research process and comes out of a process of constant interaction between data collection and analysis (Glaser and Strauss 1967; Goulding 2001). The emphasis is put on the importance of participants' behaviour as the engine of the meanings generated and interpreted when building the final theory (Glaser and Holton 2004). The main novelty of this approach, compared to other qualitative methods, is the aim to go beyond mere description to the generation of new and innovative theories, with no intention of generating universal laws (Goulding 2001).

Although the use of grounded theory in business research is generally scarce and also misunderstood (O'Reilly et al. 2012), there do already exist some significant examples of innovation research that have relied on its principles. As mentioned earlier, Van de Ven et al. (1989) referred to the approaches of Glaser and Strauss (1967) when designing research on innovation process, and made some observations about requirements for such kinds of innovation studies. These included: (1) the creation of a clear set of concepts to select and to describe the field under study; (2) the employment of systematic methods for observing the changes in the field over time; (3) the establishment of methods for representing the data obtained from the field, and for facilitating the identification of process structures; (4) the definition of a theoretical context that helps to explain the process structures identified, and to evaluate whether existing theory fits these structures.

These four requirements have been relevant for this research since they outline the necessary steps for using grounded theory to collect and analyse data about innovation processes (Nisbet 1970; Pettigrew 1985; Van de Ven et al. 1989).

It should be pointed out that in grounded theory there is no assumption of the linearity of facts, nor about the particular stages to be followed in a study. Moreover, an initial literature review is only used for providing an overall context, and as a guideline for defining the conceptual framework for the study. Time in the field alternates with periods of data analysis, such that armed with the first conclusions obtained from data analysis, the researcher turns to the field again to obtain new data, before then carrying out further analysis for which new reviews of the most suitable literature to be employed are carried out (Glaser 1978; Goulding 2001).

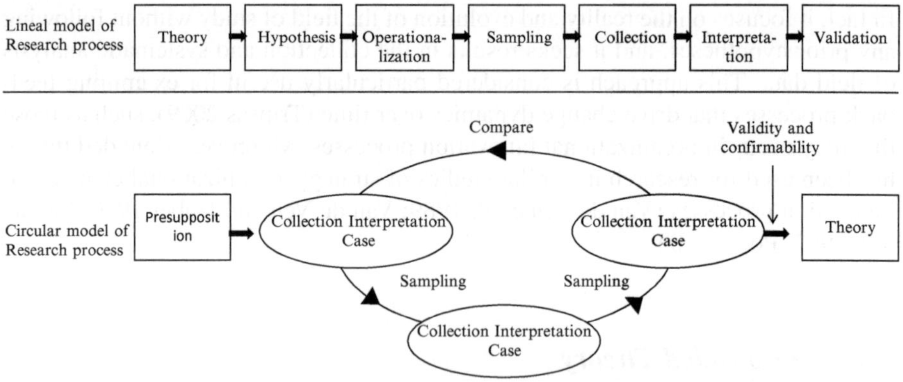

Fig. 4.1 Linear and circular models of the research process (Source: Flick 2004, p. 99)

This iterative approach, engaging in a constant comparative cycle, continues until information is saturated, so that confirmability can be checked and global conclusions made, with the overall aim of building a final theory (Trinidad et al. 2006). This process is represented by a circular model shown in Fig. 4.1.

4.3.2 *Field Settings*

This research was conducted in an industrial small and medium-sized enterprise (SME) in the Basque Country. This region offers a good context for research since there has been a growth among local companies of monitoring of organizational innovation processes. This has been promoted by the region's Competitiveness Plan 2010–2013 and its Technology, Science and Innovation Plan 2015 (Basque Government 2010, 2011).

In line with the principles of grounded theory, theoretical sampling has been applied for case election (Glaser and Strauss 1967; Eisenhardt 1989; Goulding 2001; Glaser and Holton 2004; Trinidad et al. 2006). To be included in the sample, a company had to be: (1) Basque; (2) developing an organizational innovation process on one or more dimensions (i.e. in respect of enterprise practices, work organization and/or external relationships); (3) prepared to be studied under the objectives and design of this research. Meeting with these requirements, Ennera was selected as a case study. It is a small company subsidiary of Grupo CAF, located in Ibarra (Gipuzkoa, Basque Country) and working in renewable energies and sustainable mobility.

4.3.3 *Data Collection*

Multiple data sources have been used to apply triangulation of data and conclusions (Miles and Huberman 1984) as a basis for case study validation (Yin 2009), as a way of achieving theoretical saturation using grounded theory, and as a guarantee of the quality and reliability of the research (Trinidad et al. 2006). Five data collecting methods were used for the first round of fieldwork. These were employed between February 2011 and May 2012, and addressed the entire history of the Ennera company, founded in 2007. These methods were:

1. Examination of documents. Two types of documents were collected, namely general public information (useful for case selection and the preparation of interviews), and internal enterprise documents (such as an organization chart, protocols on the use of the internet, and information on task distributions).
2. The carrying out of semi-structured interviews. Overall, 22 interviews were conducted. An initial 3 were with the General Manager, and were used to inform on case selection and to help prepare next steps. Then, 16 more were carried out during participant observation with various enterprise members, selected through a snow-ball method, until saturation was reached. 3 more interviews were carried with external partners that had been mentioned by Ennera employees. All the interviews were aimed at acquiring particular narratives of the interviewees' experiences about the organization and were also completed with structured questions related to the definition of organizational innovation.
3. The engagement in participant observation. Over 3 months (November 2011 to January 2012), for 2 days a week, the researcher immersed herself in life at Ennera, making observations from an office sharing desk, in meeting rooms, in lunch and coffee breaks with enterprise members, and during participation as an observer in internal meetings and in the December Board Director's Committee meeting. Very precise field notes were taken on a constant basis considering the physical environment, the social atmosphere, the carrying out of individual and collective activities, the occurrence of both common and special events, the various human profiles, the conversations and relationships….
4. The use of graphic evidence. Three types of such evidence were collected during the fieldwork: (1) photographs of the office; (2) photographs provided by participants (that they considered relevant to their stories); and (3) online graphic follow-up, drawing on the enterprise's web site and on social media platforms.
5. Devolution or member-checking: the first conclusions were presented to the study participants in order to observe their behaviour and reactions, for use as additional data.

4.3.4 Data Analysis and Preliminary Results

Data analysis was carried out using the constant comparative method that is associated with grounded theory (Glaser and Strauss 1967; Glaser and Holton 2004; Trinidad et al. 2006). This method was used to build theory by systematically comparing incidents identified through the data, in parallel to its collection. Figure 4.2 shows the process of the constant comparative method, outlining the basic structure of analysis followed in the research with Ennera.

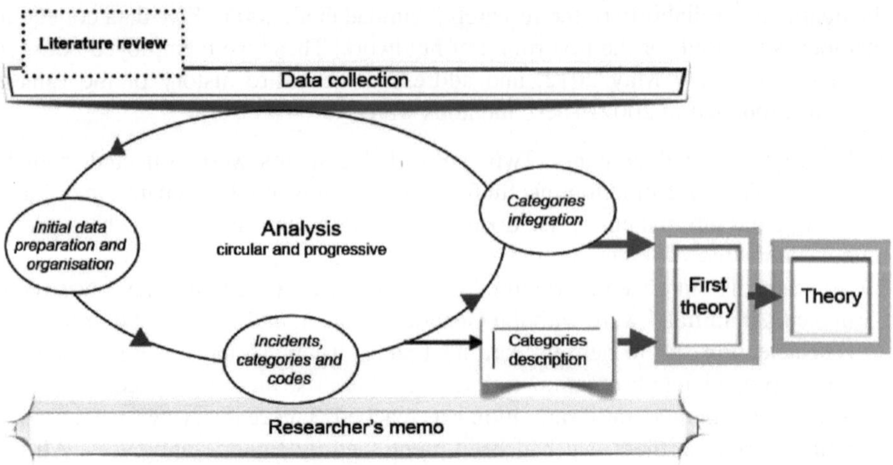

Fig. 4.2 Analysis process using the constant comparative method (Source: Own elaboration)

Initial data preparation and organization implied that, during fieldwork, all data was digitalized as audio, image or scanned documents; classified on physical and digital files; and precisely codified to represent the type of data source, the collection date and the participant (maintaining confidentiality). At each step notes were written as a researchers' memo.

Anything seen as relevant to the study were isolated as incidents. Each incident had a meaning, represented by the researcher with a category, as a classification element used for the theoretical explanation of each incident. Then, different categories were given corresponding codes (as tags), facilitating thereby analysis. This process resulted in 1,904 incidents classified into 27 categories.

Analysis began describing the meanings for each category based on the data (participants' behaviour being taken as the engine for meaning generation and interpretation), and by representing graphically each code assigned to its category (using tag clouds), then turning once again to the data to make constant comparisons.

As categories were described, connections among them were detected and represented in a timeline of events and on a conceptual map. Once description was finished, all connections in the timeline and the conceptual map were compared with the data, and main topics were identified (27 categories were organised into 10 topics).

A first step in the constant comparative method is to make sense and meaning of the relationships among the topics. In this case the timeline and conceptual map previously designed were used to connect topics, and the narrative of the case study was written down, describing a story of organizational innovation as derived from the data.

This story in fact describes the evolution of Ennera as it changed from a project based workplace organization (until 2010) to one characterised as a functionally departmentalized model (in 2011), which resulted in an enhancement of vertical decision-making processes. However, this hierarchical tendency was only partially developed, since it remained embedded in deeply rooted horizontal business practices and communication processes. In fact, for almost 5 years, all the administrative tasks (including, for example, reception, telephone answering, travel booking, lunch booking, office-kitchen cleaning, and purchasing of office material) were carried out by all the 20 members of the Ennera team, no matter what their qualifications, experience or role in the company. Similarly, news about clients, partners, projects or proposals were communicated ad hoc by anyone in the team, and people could informally get together to discuss what was happening, without needing to wait for an official meeting to be called by a Manager or the Director. In fact, the main challenge surfacing in 2012 was the need to formalize certain business practices in accordance with the new departmentalized work organization, while maintaining an informal and family-like organizational culture, that has arisen as a key element in the constant evolution of the organizational innovation process in Ennera.

The work organization method implemented was new to the organization but not new to the market; however, the business practices being developed were completely self-generated and, as a creative process occurred whereby the new business practices were combined with new commonly-known practices (such as the creation of Committee of Directors meetings), the resulting innovation could also become new to the market. However, this novelty aspect requires further fieldwork analysing the evolution from 2011 onwards.

What is evident is that this organizational innovation process does not respond to a planned process, nor neither to a purely adopted one. Neither Manager nor employees had pre-defined ideas of the organizational method to be deployed, and it is clear from the story that it is not possible to distinguish initiation and implementation stages. What has happened has been a sequence of actions and decisions that have evolved on a trial and error basis.

The main driver has been the strategic goal to gain a competitive advantage in the renewable energy market by offering a technology-based value-added service, one that is mainly differentiated from that of competitors by the mode of approaching the market: honest, transparent and coherent. This positioning is constantly reinforced and is grounded in the way staff work and behave on a daily basis.

In 2011, the year when the main organizational innovations were developed, the company conducted a corporate branding process in which all staff participated in core decisions, such as what should be the company's vision and value statement. At that time, it was easy to understand why from out of such a participatory process, held in an office with totally open information exchange, with flexible working schedules, where Management positions could not be easily identified, and where all staff had open access to the whole ERP system (except information on people's individual salaries), the three corporate pillars that were decided upon were "transparency, credibility and honesty in all energy business activities".[1]

However, when the first round of fieldwork for this research finished in January 2012, Ennera was still immersed in an organizational innovation process, consolidating some of the innovations developed, and initiating new ones in respect of business practices and external relationships.

[1] Source: www.ennera.com.

Subsequently, the researcher at a distance followed unfolding events during 2012, and then a second round of fieldwork lasting 3 months was started in November 2012. The data collected and analysed for this period will be used to advance further the constant comparative method, to engage in further confirmation of observations, and to put forward a final theory proposal.

Initial results indicate that, in line with the Oslo Manual's definition of organizational innovation (OECD/Eurostat 2005), Ennera is implementing organizational methods on business practices and work organization that are new to the company and which are driven by a clear strategic motivation. Consequently, a real concrete application corresponds to what would be expected from organization innovation theory. However, further research would be needed if we wished to make a clear differentiation from the concept of organizational change because the same evidence fits just as well a mere organizational change definition; and so the relevance of the organizational innovation concept might be questioned. It is indeed shown that when the measure of novelty required for a change to be considered an innovation is simply that of "new to the organization", then many merely formal or departmental changes might be classified as organizational innovations, even though the potential for external diffusion and transfer would be greatly limited. From a process perspective, the preliminary results show that for Ennera what happened corresponds with what might be expected in an innovation generation process. This also shows that Damanpour and Wischnevsky's (2006) differentiation between adoption and generation innovation processes is not only applicable to technological innovation but also to organizational innovation. Therefore, there could be an expansion of knowledge about organizational innovation if such a distinction was applied when studying its antecedents, drivers, obstacles and overall processes. It is expected that more longitudinal empirical research in Ennera will shed new light.

4.4 Conclusions and Implications

This study is ongoing, but preliminary results have already shown that the definition of organizational innovation proposed by the Oslo Manual in its last edition (OECD/Eurostat 2005) is a significant step towards the concept's clarification, though it is too broad to be sufficient yet. This study proposes that future research should focus on the internal benefits of organizational innovations, and also on the potential for external diffusion and the strengthening of territory-wide competitiveness.

Also, differences between innovation adoption and innovation generation have been shown in this study to be relevant to organizational innovation processes, although in the case studied further research is still needed to achieve a final theoretical explanation. Moreover, it has been found that the application of a process perspective rather than a variation perspective has been useful for deepening understanding of a complex phenomenon like organizational innovation, especially in so far as the approach enables a thorough understanding of circular sequences of events, determined by personal behaviours.

This study has limitations. With the researcher as the main research tool, there is a risk of bias. The researcher's immersion is essential for understanding the real meaning that participants attribute to contexts and to happening events, but the need to minimise the risk of bias and achieve confirmability requires the triangulation of data sources, rigour in maintaining proper field notes, and also rigour in respect of the researcher's own consciousness (Kawulich 2005; Peñaloza and Cayla 2006; Hernández Sampieri et al. 2010).

Moreover, because of the character of qualitative research methods, it is not possible to generalize results from this research, so applicability will be found facilitating transference to other contexts (Williams et al. 2005).

In conclusion, this study makes a contribution to a deeper knowledge of organizational innovation, and to opening up future research tracks about its influence on competitiveness. Besides, it expects to provide a useful research method, based on grounded theory that can be applied to the study of other innovation processes. There may also be practical implications for Basque companies and policy makers, since they are working on developing an innovation based sustainable competitiveness model, and on acquiring a thorough understanding of the important drivers of, and obstacles to, organizational innovation processes.

References

Angle HL, Van de Ven AH (2000) Suggestions for managing the innovation journey. In: Van De Ven AH, Angle HL, Poole MS (eds) Research on the management of innovation. Oxford University Press, New York

Armbruster H, Kirner E, Lay G, Szwejczewski M, Coriat B (2007) Patterns of Organisational Change in European Industry (PORCH). Ways to strengthen the empirical basis of research and policy. Fraunhofer-Institut für System- und Innovationsforschung, Karlsruhe, European Commission, Directorate-General for Enterprise and Industry, Innovation Policy Unit

Arraut Carmargo LC (2009) Innovación Organizacional Para Mejorar La Productividad Y Competitividad De Las Empresas Petroquímicas-Plásticas De Cartagena De Indias. Doctoral Thesis. Mondragon Unibertsitatea

Basque Government/Eusko Jaurlaritza (2010) Plan De Competitividad Empresarial 2010-2013. http://www.euskadinnova.net/es/portada-euskadiinnova/politica-estrategia/plan-competitividad-empresarial-2010-2013/111.aspx. Accessed 10 Feb 2013

Basque Government/Eusko Jaurlaritza (2011) Plan De Ciencia Tecnología E Innovación 2015. http://www.euskadinnova.net/es/innovacion-tecnologica/ambitos-actuacion/pcti-2015/163.aspx. Accessed 10 Feb 2013

Daft RL, Becker SW (1978) The innovative organization. Elsevier, New York. Quoted in Damanpour and Wischnevsky (2006)

Damanpour F, Wischnevsky JD (2006) Research on innovation in organizations: distinguishing innovation-generating from innovation-adopting organizations. J Eng Technol Manage 23:269–291

Damanpour F, Szabat KA, Evan WM (1989) The relationship between types of innovation and organizational performance. J Manage Stud 26(6):587–601

Dodgson M (2000) The management of technological innovation. Oxford University Press, Oxford/New York. Quoted in Som et al. (2012)

Drucker PF (1985) The discipline of innovation. Harvard Business Review 72–76

Eisenhardt KM (1989) Building theories from case study research. Acad Manage Rev 14(4):532–550
Flick U (2004) Introducción A La Investigación Cualitativa. Ed. Morata, Madrid
Glaser BG (1978) Theoretical sensitivity: advances in the methodology of grounded theory. Sociology Press, Mill Valley
Glaser BG, Holton J (2004) Remodeling grounded theory. March 2004. http://www.qualitative-research.net/index.php/fqs/article/view/607. Accessed 10 Feb 2013
Glaser BG, Straus AL (1967) The discovery of grounded theory. Strategies for qualitative research. Weidenfeld and Nicolson, London
Goulding C (2001) Grounded theory: a magical formula or a potential nightmare. Market Rev 2:21–34
Greenan N (2003) Organisational change, technology, employment and skills: an empirical study of French manufacturing. Camb J Econ 27:287–316
Hernández Sampieri R, Fernández Collado C, Baptista Lucio P (2010) Metodología de Investigación, 5th edn. McGraw-Hill, México
Kanter RM (1988) When a thousand flowers bloom: structural, collective and social conditions for organizational innovation. In: Cummings LL, Staw BM (eds) Research in organizational behavior, vol 10. JAI Press, Greenwich. pp 169–211. Quoted in Damanpour and Wischnevsky (2006)
Kawulich BB (2005) Participant observation as a data collection method. Forum Qualitative Sozialforschung/Forum Qual Soc Res 6(2):art. 43
Kirner E, Som O, Armbruster H, Lay G, Zhou W (2008) Measuring organisational innovation – concepts, indicators and outcomes. 6CP Workshop non-technical innovations. Fraunhofer Institute for Systems and Innovation Research, Karlsruhe, 16–17 Oct 2008
Kline, S. J., Rosenberg, N. (1986) An overview of innovation. In: Landau, R., Rosenberg, N., (eds.): The positive sum strategy. Harnessing technology for economic growth. National Academy Press, Washington D.C., pp. 275–305. Quoted in Som et al (2012)
Lam A (2005) Organizational innovation. In: Fagerberg J, Mowery D, Nelson RR (eds) The Oxford handbook of innovation. Oxford University Press, Oxford
Miles M, Huberman AM (1984) Qualitative data analysis. SAGE, Beverly Hills
Nisbet RA (1970) Developmentalism: a critical analysis. In: Mckinney I, Tiryakin E (eds) Theoretical sociology: perspectives and developments. Meredith, New York, pp 167–206
O'Reilly K, Paper D, Marx S (2012) Demystifying grounded theory for business research. Organ Res Methods 15(2):247–262
OECD-Eurostat (2005) Oslo manual, OECD proposed guidelines for collecting and interpreting technological innovation data, 3rd edn. OECD Publishing, Paris
Peñaloza L, Cayla J (2006) Writing pictures/taking fieldnotes: towards a more visual and material ethnographic consumer research. In: Belk RW (ed) Handbook of qualitative research methods in marketing. Edward Elgar, Cheltenham, pp 279–290
Pettigrew A (1985) The awakening giant: continuity and change in ICI. Basil Blackwell, Oxford
Poole MS (2004) Central issues in the study of change and innovation. In: Poole MS, Van De Ven A (eds) Handbook of organizational change and innovation. Oxford University Press, New York
Roberts EB (1988) Managing invention and innovation. Res Tech Manag 31:11–29. Quoted in Damanpour and Wischnevsky (2006)
Rothwell R (2003) Towards the fifth-generation innovation process. In: Jones O, Tilley F (eds) Competitive advantage in SMEs. Organising for innovation and change. John Wiley & Sons, Chichester, pp 115–135. Quoted in Som et al. (2012)
Som O, Diekmann J, Solberg E, Schricke E, Schubert T, Jung-Erceg P, Stehnken T, Daimer S (2012) Organisational and marketing innovation – promises and pitfalls? PRO INNO Europe: INNO-Grips II report. European Commission, DG Enterprise and Industry, Brussels
Tidd J, Bessant J (2009) Managing innovation. Integrating technological, market and organisational change, 4th edn. Wiley, Chichester. Quoted in Som et al. (2012)

Trinidad A, Carrero V, Soriano RM (2006) Teoría Fundamentada "Grounded Theory". La Construcción De La Teoría A Través Del Análisis Interpretacional. Centro De Investigaciones Sociológicas, Madrid

Tripsas M (2009) Technology, identity, and inertia through the lens of the digital photography company. Organ Sci 20(2):441–460

Van de Ven AH (1986) Central problems in the management of innovation. Manag Sci 32:590–607. Quoted in Damanpour and Wischnevsky (2006)

Van de Ven A (1992) Suggestions for studying strategy process: a research note. Strateg Manage J 13:169–188

Van de Ven A, Huber GP (1990) Longitudinal field research methods for studying processes of organizational change. Organ Sci 1(3):213–219

Van De Ven AH, Angle HL, Poole MS (1989) Research on the management of innovation: the Minnesota studies. Ballinger Publishing, Cambridge, MA

Williams M, Unrau Y, Grinnell R (2005) The qualitative research approach. Oxford University Press, New York

Yin RK (2009) Case study research. Design and methods, 4th edn. SAGE Publications, Thousand Oaks, CA

Chapter 5
Unfurling Organizational Innovation in Public Services: The Case of a Public Research Organization

Carlos Martin-Rios, Charles Heckscher, and Cesar Gonzalez

Abstract This paper uses a longitudinal case study of an internally driven organizational innovation and redesign process at a scientific or public research organization (PRO) as a way of illustrating innovation dynamics that result from the need to formulate a new strategic mission for the organization as a response to wider environmental and institutional pressure. Based on in-depth interviews with key participants, supplemented by a review of project reports, contract archives, publications and press coverage, this paper illustrates that organizational renewal is a complex phenomenon in PROs and that innovations in essential elements of the formal structure, work practices and values can serve as important enablers of change in highly rigid work environments; it also shows that the introduction of certain management principles borrowed from private organizations may accelerate change by providing a strong basis for developing a shared collaborative organization in public research.

5.1 Introduction

The need for continuous sustainable innovation reflects a way in which public services, such as public research organizations (hereinafter, PRO) respond to changes in the external environment. Scientific organizations or PROs are defined as government-funded research organizations that include non-profit research institutions, government agencies, and laboratories. These institutions focus on carrying out basic and, increasingly, applied research, which is all the more remarkable

C. Martin-Rios (✉) • C. Heckscher
Center for Organizational Learning & Transformation (COLT),
Rutgers University, New Jersey, USA
e-mail: cmartinrios@cminnovation.org; heckscher@charles.edu

C. Gonzalez
Instituto Ciencias del Patrimonio (CSIC), Santiago de Compostela, Spain
e-mail: cesar.gonzalez@incipit.csic.es

J.-L. Hervás-Oliver and M. Peris-Ortiz (eds.), *Management Innovation*,
Springer Proceedings in Business and Economics, DOI 10.1007/978-3-319-03134-7_5,
© Springer International Publishing Switzerland 2014

considering their capacity to collaborate with other public and private R&D institutions. Although the value of PRO has seldom been questioned, their characteristics and functioning as well as the appropriate amount of funds to be supplied have received considerable attention in the media and from the scientific community. As in the case of other public services, PROs are subject to increasingly severe scrutiny and pressure for short-term results and maximization of their research output forcing them to shift away from their traditional, bureaucratic type of organization.

Yet, as Josserand, Teo and Clegg (2006) contend, the transition of PROs and, broadly, public service organizations toward post-bureaucratic or collaborative types of organization present intrinsic difficulties, particularly those associated with the refurbishment of the organization structure or changes in attitudes, values and behaviors present in these organizations (Barzelay 1992; Harris and Wegg-Prosser 2007; Parker and Bradley 2000). With few exceptions (for example, Cruz-Castro and Sanz-Menendez 2007), the transformation in public science has gone unnoticed in the organizational literature. Research into PRO refurbishment has consisted of descriptive research. Research should thus establish a link between external pressure and adaptation choices, examining whether innovations that work on the private sphere also produce effective results in PROs. Despite its importance for competitiveness, innovation in PRO has still only received limited attention (Damanpour et al. 1989, 2009). Significant questions remain unanswered, specifically, *What role does organizational innovation play in the ability of PRO to respond to environmental and institutional pressures?*

Only through in-depth empirical analyses can we assess whether innovation initiatives really facilitate adaptation. Therefore, we set out to explore the role of organizational innovation in the transformation process of a PRO by means of a longitudinal case study. Specifically, we address a critical aspect of PROs: the search for ways to renew organizational design and practice through organizational innovation. The article begins with a review of the existing literature on PRO. We then provide an account of the pressure faced by scientific organizations in terms of organizational renewal. We discuss the methods and sources of data we have selected for the PRO case we chose for our longitudinal study, and then construct a detailed process history based on our analysis of the qualitative data collected. We discuss our results and finally suggest some areas for future research.

5.2 Theoretical Background

5.2.1 Organizational Challenges for PRO

The term PRO is used to refer to a heterogeneous group of research performing centers and institutes with varying degrees of "publicness" (Crow and Bozeman 1998). The importance of PROs is emphasized by virtually any social and economic measure chosen. Most of the OECD countries, including USA, invest heavily in

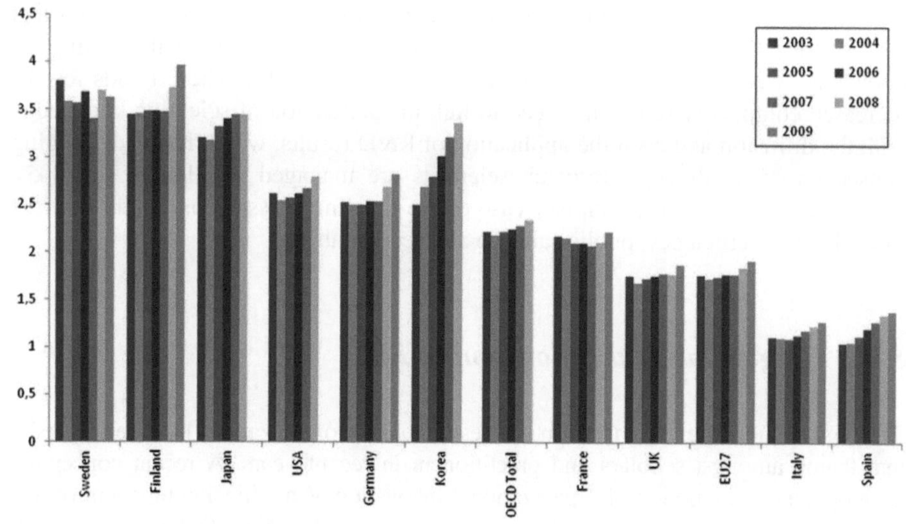

Fig. 5.1 Variation in gross domestic expenditure on R&D (percentage of GDP) for selected OECD countries (2003–2009) (Source: OECD, Main Science and Technology Indicators Database, June 2011. Note: Latest data available for Japan, USA, Korea and OECD Total is 2008)

public science. Figure 5.1 shows the variation in gross domestic expenditure on R&D (as percentage of GDP), for the period 2003–2009.

European countries such as France, Germany or Spain have built up a system for public science revolving around PROs, such as the *Centre National de la Recherche Scientifique* (CNRS) in France, the *Max Plank Gesellschaft* in Germany, and the *Consejo Superior de Investigaciones Cientificas* (CSIC) in Spain. In some way consistent with Arrow's hypothesis of under-investment in private R&D (1962), these institutions aim to provide a public research infrastructure. Publicly funded research centers are therefore established to complement research carried out at universities, foster the further development of basic and applied research, and facilitate industrial innovation and technology transfer to private industry (Albert et al. 2007; Beise and Stahl 1999; Brooks 1994).

It is commonly assumed that PROs carry out basic scientific research, while private companies are engaged in more applied efforts (Arrow 1962; Nelson 1990). Alfred Marshall in his Industry and Trade (1919) outlined the different kinds of research laboratory to establish a tripartite classification, as follows: those originators of scientific advances (i.e., PROs and publicly funded universities); those responsible for generating knowledge directed at the requirements of a particular branch of industry (i.e., R&D in private businesses); and quality control laboratories that verify that output meets the standards required. However, nowadays trends in greater economic liberalization, less public funding and greater pressure for outputs, have blurred the boundaries between the spheres (Cruz-Castro and Sanz-Menendez 2007). For PRO the traditional distinction between basic and applied science is increasingly understood in broad terms, as the level of applicability and quality of results on their research activities and funding. PROs are therefore increasingly required to become service-oriented, in the sense that they are supposed to offer products, processes and performance (Edvardsson et al. 2005; Hill 1977), while substantially relying on knowledge-work systems to achieve success (Blackler 1995). These trends toward increased complexity reflect the need to link the generation of scientific knowledge with the diffusion and even the application of R&D results, which has wide-ranging consequences on the way research scientists are managed (Mallon et al. 2005). Therefore, PROs are an exemplary case of the growing pressure on public services toward greater efficiency, public salience and accountability.

5.2.2 Organizational Innovation in PRO

The organizational renewal (reform and restructure) of organizations has been a persistent theme amongst scholars and practitioners in recent years. A recent conceptual development is the notion of organizational innovation as a critical component of this process (Armbruster et al. 2008; Birkinshaw et al 2008; Damanpour 1991). Organizational innovation (also called 'management', 'administrative' or 'social' innovation) is primarily regarded as a way in which organizations respond to environmental, market and managerial challenges by the creation, development and implementation of a new organizational method or practice that has an impact on the organization's overall success (Birkinshaw 2010; Damanpour 1991; Hipp and Grupp 2005; Lam 2005).

The organizational literature offers a variety of conceptualizations and distinctive examples of organizational innovation, suggesting a certain lack of consensus. According to Birkinshaw and colleagues, organizational innovations comprise "management practices, management processes, management techniques, and organizational structures as different facets of the rules and routines by which work gets done inside organizations" (2008, p. 828). Thus, innovations may include new internal structures, the introduction of human resource practices, changes in the web of work norms and values, or the adoption of work routines to make an organization more efficient, collaborative or professionally oriented.

Organizational innovation may potentially facilitate the development of pathways to not only change organizational objectives and structures but also non-structural aspects, such as work attitudes, values, relationships and behavior. Studies of organizational innovativeness in businesses, whose research focuses on the organizational antecedents, processes, and culture, regard innovation as a highly organic and interactive process in which the organization adapts the prevailing culture and shared values in relation to work processes and behaviors (Kanter 1988; Van de Ven and Poole 2005). In this field, the adoption and implementation process of innovations must be interpreted and reframed in accordance with the organizational context and priorities, a process that often includes active participation of leadership and various key members and groups (Birkinshaw and Mol 2006).

While there is a broad diversity of workplace structure, process and practice innovations, public services often rely on changing values as essential drivers of process innovation, as several studies have shown (Barzelay 1992; Damanpour et al.

2009; Damanpour and Schneider 2009; Harris and Wegg-Prosser 2007; Parker and Bradley 2000; Salge 2011).

Despite the theoretical arguments and empirical evidence for the impact of environmental and organizational antecedents on formal innovation initiatives, there is no clear definition of the way PROs replace the structures, practices, values and behaviors of the 'old approach' with new, post-bureaucratic and collaborative work systems.

5.3 Methodology

In order to understand under what circumstances PROs introduce organizational innovations, this paper adopted a longitudinal case study methodology (Van de Ven and Huber 1990; Yin 1994) on a government-funded research institute under the administrative authority of the Spanish government (the Ministry of Science and Innovation). Over a 4-year period between 2008 and 2011, we visited the facilities on several occasions in order to carry out in-depth interviews and collect documentary data. This setting was selected because of its potential to provide insights into the organizational innovation process of a traditional bureaucratic setting where pressure toward greater accountability and efficiency renders organizational renewal more complex. In late 2008, the organization embarked on a wide organizational innovation initiative with the aim of articulating a formal strategic mission (in terms of setting organizational objectives), addressing issues pertaining to organizational structure, culture and decision-making, and, finally, managing the organization in a more interdependent way, set apart from the bureaucratic constraints commonly found in public administration. This transformation process presented us with a unique opportunity to observe a natural experiment in internally-driven organizational innovation and redesign.

5.3.1 Data Analysis

In this paper, data were drawn from in-depth interviews, specific organization documents (a review of project reports and internal memos, training material, employee surveys and publications) and publicly available literature on the organization. The qualitative approach provided a rich and comprehensive view of the innovation initiative. Process, fine-grained qualitative data for this study involved multiple levels and units of analysis (Strauss and Corbin 1998). On the one hand, data collected at system level were mainly obtained through the Institute leader, senior management (researchers and specialists on the initiative) as well as researchers responsible for certain areas of the organization (e.g. head of the publications department or project leaders in certain fields of specialization). They provided valuable information to examine the sequence of events over time and how organizational innovations unfolded, and to identify process patterns. On the other hand, experiences regarding the change initiative were obtained through interviews with the organizational members themselves.

Information collected from interviews was obtained by interviewing 12 key participants on-site at Incipit. We interviewed all of the members participating in innovation initiatives, together with several representative senior and junior scientists, in order to determine the level of engagement achieved. Additionally, numerous informal conversations took place over the 4-year period of fieldwork.

All requests for interviews were answered positively and nobody declined to participate. Participation was voluntary. Interviewees were assured of the anonymity and confidentiality of their responses. Interviews ranged roughly from 60 to 90 min. They were digitally recorded, with permission, and transcriptions were made of all relevant portions. A semi-structured interview template was used to guide the research, but the template was not followed strictly, and interviewees were prompted to talk freely about whatever seemed important to them.

Data from the interviews were used for two purposes. First, a subset of initial interview accounts served to create an outline of the critical events, helping to trace the change pathway followed by the institute. Secondary sources of information, like internal reports, were then used to validate the details of the proposed timeline. The second round of interviews was then used to probe the innovation practices in place and understand why these particular innovations were taken at different times. Based on the data gathered and analyzed in the earlier stages of the research, further data was extrapolated and developed so as to reflect a more exhaustive and in depth review of the newly gathered data.

Data analysis was performed to trace the chronological sequence of activities that occurred throughout the organizational renewal period. Using our theoretical foundation of organizational innovation and the development of collaborative scientific organizations, we conducted our event history analysis around a thematic analysis to identify how organizational innovation and the broad change initiative emerged and developed over time. Additional data on quantifiable outputs (e.g. publications, participation in research projects, etc.) was analyzed in terms of how innovations helped the organization to fulfill its goals. Also, each interview was analyzed to identify and explore themes, understandings and perceptions relating to the change initiative and the organizational innovations launched between 2008 and 2011. These data were analyzed via the open-coding procedure, grouped into secondary or axial codes and represented here as themes. In this way we were able to identify broader key themes within the organizational change initiative.

5.4 Findings

5.4.1 Antecedents

The Spanish National Research Council is a government-funded research organization under the administrative authority of the Spanish government (Ministry of Science and Innovation). It is formally organized as a collection of over 130 research institutes, each of which specializes in a particular area of knowledge.

Institutes may be regular CSIC institutes or hybrid institutes between CSIC and a second partner, such as a university or a regional government. Each Institute may choose to create an internal organization of distinct laboratories or research units.

Our research case involves the Institute of Heritage Sciences (Incipit, in its Spanish acronym), a regular (i.e. non-hybrid) small institute of over 45 people engaged in basic, non-applied research on the field of heritage. Incipit is the youngest institute of the CSIC; in fact, it was still formally in the process of creation at the time of writing. When the study described in this paper began, the people, resources and endeavors that today are part of Incipit were then part of The Heritage Laboratory (LaPa in its Spanish acronym), dependent on another CSIC Institute and partnered with the regional government of Galicia. In January 2010, after being split off from the Institute to which it belonged, Incipit was formally created, and all of the staff and resources assigned to LaPa so far were transferred to it. The setting up of Incipit was due to three major achievements: Firstly, outstanding performance measured in terms of scientific output (publications, research projects granted, conferences attended, postdoc researchers hosted, Ph.D. theses written under their supervision, etc.); secondly, the recognition of its capacity for self-management; thirdly, good performance in resource allocation and expenditure; and finally, significant human capital growth (currently close to 50 people working both full time and part-time) and research areas covered (around 10 areas of knowledge).

In practice, Incipit hosts specialists (at least one full time scientist) in a variety of fields, including archaeology, anthropology, geology, soil sciences, astronomy, geography, cultural sociology, art history, architecture, and information technologies. Today, Incipit has broadly focused on cultural heritage in a way that reflects the richness and diversity of the field, creating opportunities for cross-disciplinary research. For example, apart from traditional archaeological researchers, it would willingly incorporate researchers working on heritage as a marketing tool for tourist services, or exploring heritage as a political instrument.

Although the innovation initiative was launched before Incipit was formally created, the new organizational status gave researchers a much greater degree of autonomy, particularly in the choice of mission objectives and organizational design, which soon crystallized into the need to formulate a formal collaborative, multidisciplinary strategy. The broadening of academic research areas to include a multilevel, multidisciplinary approach would enhance scientific results by creating a space for new ways of collaboration and learning partnerships. Multidisciplinary work requires a new strategic vision, but also attitudes, values and behaviors (Jackson 1996), which clash with most standard notions of work in bureaucratic organizations and therefore calls for new organizational arrangements. However, public research policies and organization in Spain (and Europe) are strongly based on the concept of knowledge areas or disciplines, and although the idea of interdisciplinarity is often praised, an actual interdisciplinary implementation is unlikely to survive in an environment built for traditional, discipline-oriented institutes.

Since late 2008, Incipit has gone through an intense restructuring process with the twofold aim of establishing an organizational structure that overcomes the rigid nature of the former bureaucratic design, and fostering an organizational infrastructure capable of changing values and behaviors over time by relying on a collaborative culture that allows for both innovativeness and new interdisciplinary collaborations. Both initiatives pervade the entire organization, from the most senior staff through to young, newly incorporated PhD students and assistants. In parallel, to help people in this endeavor, Incipit formulated a shared organizational mission in order to become a more service-oriented organization focused on addressing the needs and demands of multiple stakeholders. An extensive communication program that encouraged, facilitated, and supported involvement in decision-making was also set underway. For the most part, the organization has been successful in formulating objectives and constructing channels of communication for its interchanges with employees.

5.4.2 *Organizational Innovation Initiatives*

We have constructed a process history of the 'innovation pathway' taken by Incipit around three major stages: first, during the 'bureaucratization phase' the organization revolved around informal work relationships integrated in work practices connected with the traditional bureaucratic form of most scientific organizations. This gave way to the 'innovation phase' characterized by several organizational innovations mainly aimed at delivering an organization mission and developing a new organizational structure. Finally, during the 'consolidation phase' the Institute is expected to consolidate the implementation of these innovations in the coming years by promoting a collaborative culture that fosters interdisciplinary team arrangements and greater individual accountability towards a common purpose. Figure 5.2 illustrates the stages of the change initiative according to the organizational characteristics (in terms of structure and resulting culture) and the innovations implemented.

In terms of formal organization, Incipit's work values were shaped by its own founder. Decisions were centralized at the leader level. He supervised what to research, how to operate, or who would be in charge of what. Beyond that, the research team operated in a rather informal way. This type of relaxed, informal culture of 'free agents' offered advantages to self-motivated researchers, such as allowing disciplinary autonomy from outside control; it also limited the capacity to organize and carry out complex projects independently. This lack of decision-making ability slowed down the organization, making it highly dependent on the leader, at the risk of becoming the bottleneck in the decision-making chain. As one senior researcher put it,

> There are people who could not make a decision or just didn't want to make it. This overexposed [the director]. They came to him for every decision and expected him to respond. Even worse, [the director] seemed to like having to respond to every request.

Five years after its inception, Incipit rapidly grew in size and scope to the point of where a new 'organization' was needed on both levels: its formal structure, and its values and behaviors. The previous informal organization of work had given way to an increasing number of rules and procedures. This is reflected in the following excerpt from one senior researcher:

> Coordination was becoming unattainable. For example, at a certain point we didn't even know who to go to in order to take time off for vacation. A standard procedure was in place;

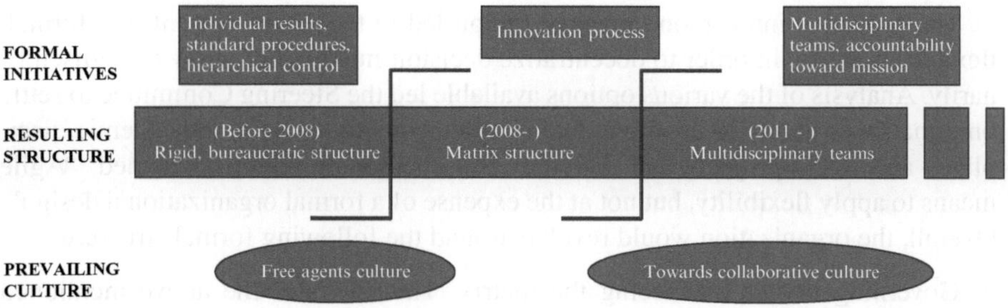

Fig. 5.2 Stages in the innovation process at Incipit

> however, someone would tell you, 'you can send an e-mail to everyone', others would say 'go and ask [the director]', or you even would hear something like, 'just leave but let [the director] know'. Eventually, people were on and off with little or no control whatsoever.

Since direct command from the headquarters of the CSIC, as the governing institution, did not interfere to any great extent, each institute was allowed to operate and organize itself quite independently, with a few exceptions. There were some formal procedures that applied to every Institute, and which were centrally dictated. For example, institutes did not have any say in formal human resource practices—such as recruitment, compensation, and promotion procedures—which are the same procedures in place for the entire public administration in Spain. The national collective bargaining agreements for public officials applied to every employee in the public system, but did not extend to PhD students, postdocs, visiting scientists, and contingent workers. Otherwise, Incipit could organize its work practices as it saw fit.

The prospects of a transformation increased when the organizational leaders set up a series of meetings to discuss the future of the Institute. Two governing bodies, the Scientific Board and Steering Committee, were established to design and oversee an action proposal or 'innovation pathway'. Members of the Steering Committee engaged in a review of considerable body of academic literature, as well as public reports and documentation from service consulting firms. Moreover, all employees and researchers were given the opportunity, in several forms, to let their views be known, and to participate in the transformation initiative. They also invited several external experts on organizational design and transformation to discuss their opinions on this process. Finally, they carried out an informal benchmarking of what other R&D institutes had done before in terms of organization.

The common approach is to connect research strategies with the leader's scientific field of expertise (e.g. a renowned biologist expert on white cells manages her institute, set up lines of research, and hires and manages scientists on related fields to work for her). This 'leader-centered' approach was rejected on the grounds that cultural heritage was naturally a fertile, transversal research field. As the director pointed out,

> Often, research institutes are just an amalgamation of several independent research teams working for a leader. My goal is that our Institute be a group of teams working interdependently to master different aspects of heritage and when necessary, rally together to offer 'one-single view' of the Institute.

The proactive innovation moves at Incipit led to the establishment of a formal, flexible structure, in order to decentralize decision-making and grow multidisciplinarily. Analysis of the various options available led the Steering Committee to settle on a matrix structure to ensure a dual simultaneous focus on both academic disciplines and R&D projects and activities. As one senior scientist claimed: "Agile means to apply flexibility, but not at the expense of a formal organizational design". Overall, the organization would revolve around the following formal structure:

a. Governing bodies overseeing the matrix. This includes the above mentioned Scientific Board and Steering Committee.
b. On one axis of the matrix, a number of subjects were identified (one per academic discipline or area of specialization), each with its own coordinator. Subjects were conceived as professional communities that share a common body of knowledge and/or methodological approach.
c. On the other axis of the matrix, R&D endeavors (such as research projects or activities), each with its project manager. Projects and activities were conceived as managed efforts in time, with well-defined purposes and allocated teams and resources. An additional governing body called the Coordinators Board was created, which brings together all of the subject coordinators to resolve day-to-day issues connected with ongoing projects and the settlement of disputes.
d. Supporting the matrix, several horizontal services (such as library services, administration, training, health and safety, coordination of visiting researchers, spin-off hatchery, or information technology services), each managed by one volunteer researcher or specially hired technicians.

Moving towards a new way of working called for a fundamental reappraisal of the roles of everyone involved. At the same time, two specific decision-making mechanisms were designed on top of this structure. On the one hand, tactical decisions at the project level would be made by project managers with as much autonomy as possible, avoiding the continuous flow of checks and approvals with upper management that had been usual until that point. On the other hand, strategic decisions would be made by consensus within designated groups, such as the governing bodies or each of the subjects. Moving towards these decision-making mechanisms entailed a significant amount of training and re-conceptualizing work dynamics by most of the staff at Incipit.

5.4.3 Innovating Through Organizational Culture

The content of the efforts focused on bringing about change resulted in a renewed interest in addressing the organizational culture, and how to evaluate the broad contribution of a collaborative culture. The Scientific Board defined the following values associated with their work:

- Celebrating excellence.
- Commitment to accountability and collaborative decision-making.
- Operating an agile but formal way of working.

It became evident from the interviews that one reason for promoting a collaborative culture was to place emphasis on the control logic that would govern the organization. The important concern underlying this innovation was the need to address the central paradox of control and autonomy in scientific organizations. As explained above, its importance is related to the very characteristics of PROs themselves, which render traditional, bureaucratic forms of organization less appropriate. Thus, whereas coordination and control in the traditional hierarchical firm revolves around standard procedures and reliability-focused practices, in the resulting organization it was expected to be based more on exploration and on learning-focused practices. As the Institute developed a collaborative infrastructure, the logic of command-and-control seemed ill fitted to organizational requirements such as collaboration, dealing with uncertainty or embracing multiple decision-making variability. Work values related to a shared orientation were particularly important in the new structure where members must necessarily exercise a great deal of discretion. This is consistent with findings in previous studies on the topic (Simpson and Powell 1999). Scholars have made important contributions in this regard by specifying the impact that these differences should have on work norms and values (Simons 2005).

The objectives of this process change were to promote a collaborative culture by means of fostering: (a) information channels to establish 'connectedness' across all researchers; (b) interdisciplinary and collaborative work teams; and (c) resource mobilization to operate in a multilateral fashion. For example, the open sharing of information was perceived in this case as an advantage. Incipit has implemented a sophisticated communication system to strengthen the collaborative culture in place. In the broad sense of 'connectedness', the so-called process of multimode sharing of information was deemed appropriate to uphold the newly developed strategic mission.

Innovation initiatives resulted in establishing and maintaining cross-level, interdisciplinary work teams, in a way that 'accord less value to "doing a good job" or "achieving the defined objectives"; they accord highest praise for people who are able to look beyond their specific roles and who do whatever is needed to advance the common purpose' as stated by Adler and Heckscher (2011, p. 12). These innovations served the purpose of reorienting toward collaborative values in order to place greater emphasis on multidisciplinary work, common purpose, efficiency and outcomes, or what Simpson and Powell call a 'multiple project archetype' (1999). For these reasons, organizational transformation toward multidisciplinary work was deemed especially important.

5.5 Conclusions

This study was aimed at untangling the pursuit of collaborative organization in one knowledge-intensive PRO by means of the organizational innovation of its formal structure and organizational culture. We believe that our research findings are important for a number of reasons. Firstly, the association between public service organizations and organizational innovation is important. Existing research on radical organizational transformation toward increasing professionalization of services has focused on private firms, yet radical new organizational design occurs more frequently in private firms than in public organizations (Greenwood and Hinings 1996; McNulty and Ferlie 2004). We contend that organizational innovations borrowed from private firms can serve as an important enabler of adaptation in highly rigid work environments, as often characterized in PRO.

Organizational innovation is more suitable for public organizations, and is therefore more commonly adopted by them. To do so, innovations redesign the formal structures found in the bureaucratic form to achieve interdependence and collaboration, rather than hierarchy and fragmentation. Moreover, because values and behaviors are embedded in roles, rather than intervening in formal structures and expecting that work patterns will change in response, a mutually reinforcing model of organizational innovation would suggest the need to intervene in both formal structures and work values to enable a PRO to break away from existing patterns and respond to new conditions (e.g. Damanpour et al. 2009; Harrow and Willcocks 1990).

Collaborative organization may represent an answer to the increasing need for continuous sustainable innovation and adaptability in PROs. The notion of collaborative culture represents a departure from norms and values in traditional, bureaucratic settings (Heckscher 2007). It addresses the diverse pool of knowledge, skills, and experience brought by people from different spheres to the collaborative effort. It is suggested that organizations pursuing a collaborative culture place strong emphasis on aligning the standard bureaucratic interest of individual performance with the need to build collaborative work relationships and shared professional values as they seek to adopt greater emphasis on common purpose and outcomes in order to thrive (Clanon 1999; Hansen and Nohria 2004). Collaborative organizations seek timely integration, the sharing of knowledge and effective collaboration across organizational levels. In this sense, individual effort is fused with group outcomes derived from collaborative efforts. It requires the active involvement of all of the participants in the work process, which must be accompanied by reorganizing the workplace, and this is expected to become a continuous innovative process, and no longer a series of discrete events.

In summary, with this study we contribute to the growing body of knowledge analyzing the relationship between public sector innovation and organizational success by means of establishing new organizational objectives that are closer to societal demands for greater sustainable innovation and adaptability, and a continuous focus on work values promoting responsiveness and accountability. A commitment to accountability and collaborative decision-making can serve as an important enabler of change in highly rigid work environments. This paper constructs a detailed process history of the 'innovation pathway' taken by the organization to show certain evidence that the introduction of these types of innovation initiatives may accelerate change by providing a strong basis for collaborative public research organizations.

References

Adler P, Heckscher C (2011) Collaborative community is the basis of organizational ambidexterity. Unpublished working paper

Albert A, Granadino B, Plaza L (2007) Scientific and technological performance evaluation of the Spanish Council for Scientific Research (CSIC) in the field of biotechnology. Scientometrics 70:41–51

Armbruster H, Bikfalvi A, Kinkel S, Lay G (2008) Organizational innovation: the challenge of measuring non-technical innovation in large-scale surveys. Technovation 28:644–657

Arrow KJ (1962) Economic welfare and the allocation of resources for invention. In: Nelson R (ed) The rate and direction of inventive activity. Princeton University Press, Princeton, pp 609–636

Barzelay M (1992) Breaking through bureaucracy: a new vision for managing in government. University of California Press, Berkeley

Beise M, Stahl H (1999) Public research and industrial innovations in Germany. Res Policy 28(4):397–422

Birkinshaw J (2010) Reinventing management: smarter choices for getting work done. Wiley, San Francisco

Birkinshaw J, Mol M (2006) How management innovation happens. Sloan Manage Rev 47:81–88

Birkinshaw J, Hamel G, Mol M (2008) Management innovation. Acad Manage Rev 33:825–845

Blackler F (1995) Knowledge, knowledge work and organizations: an overview and interpretation. Organ Stud 16:1021–1046

Brooks H (1994) The relationship between science and technology. Res Policy 23:477–486

Clanon J (1999) Organizational transformation from the inside out: reinventing the MIT Center for Organizational Learning. Learn Organ 6(4):147–156

Crow M, Bozeman B (1998) Limited by design: R&D Laboratories in the US National Innovation System. Columbia University Press, New York

Cruz-Castro L, Sanz-Menendez L (2007) New legitimation models and the transformation of the public research organizational field. Int Stud Manage Organ 37:27–52

Damanpour F (1991) Organizational innovation: a meta-analysis of effects of determinants and moderators. Acad Manage J 34(3):555–590

Damanpour F, Schneider M (2009) Characteristics of innovations and innovation adoption in public organizations: assessing the role of managers. J Public Adm Res Theory 19:495–522

Damanpour F, Szabat KA, Evan WM (1989) The relationship between types of innovation and organizational performance. J Manage Stud 26:587–602

Damanpour F, Walker R, Avellaneda C (2009) Combinative effects of innovation types and organizational performance: a longitudinal study of service organizations. J Manage Stud 46(4):650–675

Edvardsson B, Gustafsson A, Roos I (2005) Service portraits in service research: a critical review. Int J Serv Ind Manage 16(1):107–121

Greenwood R, Hinings CR (1996) Understanding radical organizational change: bringing together the old and the new institutionalism. Acad Manage Rev 21:1022–1054

Hansen M, Nohria N (2004) How to build collaborative advantage. MIT Sloan Manage Rev 46(1):22–30

Harris M, Wegg-Prosser V (2007) Post-bureaucracy and the politics of forgetting: the management of change at the BBC, 1991–2002. J Organ Change Manage 20(3):290–303

Harrow J, Willcocks L (1990) Public services management: activities, initiatives and limits to learning. J Manage Stud 27(3):281–304

Heckscher C (2007) The collaborative enterprise. Yale University Press, New Haven

Hill TP (1977) On goods and services. Rev Income Wealth 23:315–318

Hipp C, Grupp H (2005) Innovation in the service sector: the demand for service-specific innovation measurement concepts and typologies. Res Policy 34:517–535

Jackson SE (1996) The consequences of diversity in multi-disciplinary work teams. In: West MA (ed) Handbook of work group psychology. Wiley, Chichester, pp 53–76

Josserand E, Teo S, Clegg S (2006) From bureaucratic to post-bureaucratic: the difficulties of transition. J Organ Change Manage 19(1):54–64

Kanter R (1988) When a thousand flowers bloom: structural, collective and social conditions for innovation in organization. In: Staw BM, Cummings LL (eds) Research in organizational behavior. JAI Press, Greenwich, pp 169–211

Lam A (2005) Organizational innovation. In: Fagarberg J, Mowery D, Nelson R (eds) The Oxford handbook of innovations. Oxford University Press, Oxford, pp 115–147

Mallon M, Duberley J, Cohen L (2005) Careers in public sector science: orientations and implications. R&D Manage 35:395–407

Marshall A (1919) Industry and trade. A study of industrial technique and business organization; and of their influences on the conditions of various classes and nations. MacMillan, London

McNulty T, Ferlie E (2004) Process transformation: limitations to radical organizational change within public service organizations. Organ Stud 25:1389–1412

Nelson RR (1990) Capitalism as an engine of progress. Res Policy 19:193–214

Parker R, Bradley L (2000) Organizational culture in the public sector: evidence from six organizations. Int J Public Sect Manage 13(2):125–141

Salge T (2011) A behavioral model of innovative search: evidence from public hospital services. J Public Adm Res Theory 21(1):181–210

Simons R (2005) Levers of organization design. Harvard Business School Press, Boston

Simpson B, Powell M (1999) Designing research organizations for science innovation. Long Range Plann 32:441–451

Strauss A, Corbin J (1998) Basics of qualitative research. Sage, Thousand Oaks

Van de Ven A, Huber GP (1990) Longitudinal field research methods for studying the process of innovation. Organ Sci 1:213–219

Van de Ven A, Poole MS (2005) Alternative approaches for studying organizational change. Organ Stud 26:1377–1404

Yin RK (1994) Case study research: design and methods. Sage, Newbury Park

Chapter 6
Managing Risk-Taking to Enhance Innovation in Organizations

Oscar Llopis, Ana García-Granero, Anabel Fernández-Mesa, and Joaquín Alegre

Abstract Scholars have proposed that taking risks is a key factor to explain innovation performance in organizations. However, the relationship between risk taking and innovation performance is generally examined from two unconnected perspectives. From a managerial perspective, entrepreneurial orientation and leadership theories are invoked to justify a positive relationship between risk taking and innovation. From an employees' viewpoint, creativity theories suggest that a risk-taking climate influences innovative and creative behaviors. This study examines the possibility of a connection between managers' risk-taking propensity, employees' risk-taking climate, and innovation performance from both a theoretical and an empirical point of view. To do so, we draw on a dataset of 182 firms from the Spanish and Italian ceramic tile industry.

6.1 Introduction

The ability of firms to innovate is a primary factor in achieving and sustaining competitive advantage (Nelson and Winter 1982). Hence, it is widely believed that innovative behaviors should be strongly encouraged across all levels of the organization given that such behaviors are likely to exert a positive influence on organizational effectiveness (Amabile et al. 2005). The focal point of our research is the relationship between risk-taking and innovation performance from a managerial

O. Llopis (✉) • A. García-Granero • A. Fernández-Mesa
INGENIO (CSIC, Universitat Politècnica de València), Valencia, Spain

Faculty of Economics, Business Department, University of Valencia, Valencia, Spain
e-mail: osllocor@ingenio.upv.es; angargr2@ingenio.upv.es; Anabel.fernandez@uv.es

J. Alegre
Faculty of Economics, Business Department, University of Valencia, Valencia, Spain
e-mail: Joaquin.alegre@uv.es

J.-L. Hervás-Oliver and M. Peris-Ortiz (eds.), *Management Innovation*,
Springer Proceedings in Business and Economics, DOI 10.1007/978-3-319-03134-7_6,

© Springer International Publishing Switzerland 2014

and an employee perspective. The relationship between risk-taking and innovation performance is particularly fruitful. Substantial research from diverse fields suggests a close link between risk-taking and innovative behaviors in organizational settings (March and Shapira 1987). Risk-taking and innovation are intertwined due to the nature of creative behaviors in organizations.

From a managerial perspective, the link between risk-taking and innovation performance has been examined using a wide range of approaches, such as entrepreneurial orientation and leadership literatures (Covin and Slevin 1986; Ling et al. 2008; Wu et al. 2005). Risk-taking involves the investment of significant resources in activities with significant possibility of failure, which includes incurring heavy debt or making large resource commitments in the hope of reaping potential high benefits (Fernández-Mesa et al. 2012; Lumpkin and Dess 1996). Managers vary in their individual propensity to take risks. This is not trivial given that the evidence shows that a manager's preference for a risky behavior is positively associated with the attainment of higher innovation results (Ling et al. 2008). Thinking "outside the box" entails a great deal of uncertainty, and bold decisions and actions are often necessary to achieve innovative results. This implies that, compared to risk-averse managers, managers with a higher preference for risk will be more likely to consider the potential gains from risky decisions (Ling et al. 2008; Wu 2008) In March and Shapira (1987, p. 1408) words, "risk-taking is valued, treated as essential to innovation and success".

The literature on creativity provides a different, yet related, view of this relation, being more focused on the personal and contextual factors explaining why employees engage in innovative activities (Amabile et al. 1996). A fundamental idea is that creative behaviors are about challenging the status quo of given aspect of the organization. From the employee's point of view, the consequences of such challenges are uncertain. In fact, those employees displaying innovative behaviors may face negative consequences if they fail (Zhou and George 2001). For instance, Janssen (2003) demonstrates that innovative employees are likely to come into conflict with co-workers because the worker promoting a new idea is challenging established courses of action and the assumptions of co-workers. It is likely that resistance, in the form of work conflicts, will arise.

Although work from both views has significantly advanced our understanding of the nature of the link between risk-taking and innovative performance, observation of this relation through a combined lens is lacking. We believe that it would be more informative to explore the relationship between risk-taking and innovation performance at different levels of the organization. We would argue that managers' risk-taking behavior not only exerts a direct effect on innovation performance but also that the organizational risk-taking climate benefits due to a positive signaling effect deriving from managers' risk-taking attitudes.

The paper is structured as follows. First, we provide a brief theoretical review of innovation in organizational contexts. Second, we introduce the relevance of managers' and employees' risk-taking for fostering organizational innovation. In the third section, we present the conceptual model and develop our hypotheses. The last two sections test our model on a sample of 182 companies for the Spanish and Italian ceramics sector, and present our results, findings, limitations and some managerial implications.

6.2 Theoretical Framework

6.2.1 Innovation Performance

For firms innovation is central to achieving sustained competitive advantage (Teece et al. 1997). The evolution of an increasingly complex environment has made innovation an unavoidable option in plans to increase the performance, continuing growth and survival of firms (Daellenbach et al. 1999). Innovation can be defined as the successful implementation of new ideas (Amabile et al. 1996). This understanding includes novelty and usability as two indispensable conditions. Thus, innovation requires new ways to solve problems and achievement of commercial success.

Innovations can be either product or process innovations (Martínez-Ros and Labeaga 2009; OECD 2005). Product innovation is understood as a product or service introduced to meet the needs of the market or an external user; process innovation is understood as a new element introduced into production operations or functions (Damanpour and Gopalakrishnan 2001). Both types of innovation are closely related, and although firms may be more focused on product innovation, process innovation may be necessary for the successful implementation of their new products (Martínez-Ros and Labeaga 2009).

Although significant efforts have been invested in trying to understand the factors underlying innovation performance, the process carries high failure rates (Wu et al. 2005). Despite the difficulties involved in producing innovation, it is one of the main drivers of organizational growth, thus it is important to have a more fine-grained understanding of its determinants.

6.2.2 Managers' Risk-Taking Propensity

The determinants of innovation include exogenous factors such as the firm's external environment, and more malleable aspects such as the organizational culture, structure, and strategy (Papadakis et al. 1998; Vega-Jurado et al. 2008). In particular, leaders have been repeatedly recognized as strategic decision makers able to identify opportunities and make the right decisions to encourage innovation (Alexiev et al. 2010; Elenkov et al. 2005). Firms' managers involved in decision making are faced with the uncertainty intrinsic to innovation activities. Innovation needs investments of time, effort, and resources, such as increases in R&D expenses and greater allocation of management attention, although the distribution of the returns from these investments is unknown (Ling et al. 2008; Wu et al. 2005). This uncertainty and the significant possibilities of failure often lead to risk adverse behaviors and under-investment in innovation (Finkelstein 1992; Wu 2008). However, expectations of potentially high returns drive many managers to opt for risky solutions and to focus on the potential benefits of innovation rather than the potential losses (Ling et al. 2008).

Several streams of research propose that managers' risk-taking propensity can make a difference in defining the ability of firms to innovate. For instance, the entrepreneurial orientation literature conceptualizes risk-taking as one of the dimensions affecting the firm's strategic position, that is, the extent to which top managers are inclined to take business related risks (Covin and Slevin 1986). Scholars in this tradition generally focus on how an entrepreneurial orientation heightens performance (Madsen 2007; Zahra and Covin 1995).

Scholars using the upper echelon perspective study risk-taking propensity in managers and top management teams according to characteristics such as tenure and age, and their effects on innovation performance (Bantel and Jackson 1989; Liu et al. 2012; Wu et al. 2005) Work in the leadership literature assesses more directly how the propensity of top management teams for risk-taking influences performance (Papadakis et al. 1998; Peterson et al. 2003), and specifically innovative processes and outcomes (Ling et al. 2008). In general, results confirm that managers biased towards risk-taking behaviors are more likely to obtain better innovation results.

Although managers' risk-taking propensity appears pivotal for explaining innovation performance in organizations, the mechanisms linking it to organizational innovation performance remain unclear. Contextual factors in the organization may play a significant role.

6.2.3 Risk-Taking Climate

Although there are several ways to approach the different contextual features of organizations, researchers often use the notion of organizational climate to assess the social features of workplaces that facilitate or inhibit certain behaviors (Schneider and Reichers 1983). The organizational climate is a multidimensional construct that encompasses a wide range of organizational realities (James and McIntyre 1996). According to Denison (1996), organizational climate concerns those aspects of the social environment perceived by organizational members.

The concept of organizational climate has become prominent in management studies, and has been deconstructed into specific dimensions (Schneider and Reichers 1983), depending on the phenomenon under study. For instance, climate scholars have developed a construct to measure climates for justice (Naumann and Bennett 2000), creativity (Gilson and Shalley 2004), and innovation (Anderson and West 1998; Pirola-Merlo and Mann 2004), among others. Many of these "climates" occur simultaneously in an organization (Kuenzi and Schminke 2009), and measure different realities of the organizational environment. Employees conceive the climate of the organization as the source of cues about how to behave. For instance, (Gilson and Shalley 2004) found that team members who were more engaged in the creative process reported their team climate being more supportive of creativity.

A particular facet of the organizational climate that is likely to influence employees' innovative performance is the firms' risk-taking climate. Employees fear failure

(Zhou and George 2001), and innovating in an organizational setting can be viewed as risky behavior. Risk-taking means uncertainty about the potential outcomes of one's decision (Sitkin and Pablo 1992). This is a barrier that can be scaled if employees perceive that the organizational climate supports risk-taking and innovative behaviors.

6.3 Hypotheses

6.3.1 Managers' Risk-Taking Propensity and Innovation Performance

Scholars quite widely assume that the strategies of top managers chime with the organizational level aims, and that top managers' personalities and behaviors have a direct influence on organizational outcomes (Alexiev et al. 2010; Wu et al. 2005). In this sense, the actions of managers regarding risk-taking are likely to have a considerable influence over the firm' innovation performance. In this section we propose a series of mechanisms by which leaders' risk-taking propensity can influence the firms' innovation performance.

First, research on organizational behavior indicates that managers' behaviors are a powerful communicating mechanism for the whole organization (Ashkanasy et al. 2000; Grojean et al. 2004). Managers' behaviors are taken as models of appropriate behaviors in particular situations. According to social cognitive theory (Bandura 1986), individuals have the capacity to learn vicariously. Vicarious learning refers to the process of learning by observing the behavior of others and its consequences (Bandura 2001). For instance, House and Shamir (1993) suggest that vicarious learning is an important mechanism through which the values of the organization are transmitted from managers to employees. We extend this rationale to argue that those managers more prone to take risks in their organizational decisions will have a notorious influence over the rest of the organization's innovation performance. As a consequence, the firm will show higher levels of innovation performance, compared to those firms whose managers are more adverse to take risks in their managerial decisions.

Managers' risk taking behavior may spread throughout the whole organization due to the effects suggested by the signaling theory (Spence 1973). Signaling theory refers to behaviors that convey information about an individual's intentions and abilities. Management scholars have applied signaling theory to argue that, in organizations, managers are powerful signalers of desirable behaviors (Connelly et al. 2011). The main rationale for signaling theory is information asymmetry. Employees may not have full information about how they are expected to behave in particular situations (e.g. taking a risky decision versus being conservative). In order to reduce information asymmetry, managers may consciously decide to emit signals to observers. In the particular case of risk-taking, managers' risk-taking propensity may be a powerful signal to stress the importance of risk-taking behaviors among the rest of the firm. Signal receivers (here, employees), will use these signals to make more informed decisions (Cohen and Dean 2005). Taken together, the above arguments allow us to propose the following hypothesis:

Hypothesis 1: There is a positive and significant relationship between managerial risk-taking propensity and innovation performance.

6.3.2 Risk-Taking Climate and Innovation Performance

Research on creativity and innovation indicates that creative efforts require substantial investment of time and energy on the part of the individual (Redmond et al. 1993). The ultimate decision to engage in innovative behaviors belongs to the employee, and willingness and motivation to do so may be influenced by a number of organizational characteristics (Chen and Huang 2009). According to (Yuan and Woodman 2010), innovative behavior is defined as "as an employee's intentional introduction or application of new ideas, products, processes, and procedures to his or her work role, work unit, or organization". Employees deciding to search for and apply new technologies for their daily work, or suggest new ways to achieve objectives in their organization, are examples of such behaviors. These types of behaviors are likely to exert a positive effect on the organizations' overall innovation performance.

However, innovative behaviors are closely linked to risk-taking. Engaging in innovative behavior requires feeling comfortable with taking risks or at least the ability to tolerate a degree of risk. Employees may lack the motivation to take risks in their organizations for a number of reasons. Given that employees' actions are guided by expectations about the consequences of their behaviors (Vroom 1964), the perceived costs of introducing a new idea or procedure may overshadow its potential benefits. Among those costs, challenging the organizational "status quo" is prominent. Implementing or suggesting a novel procedure or idea means that existing ones are challenged. Organizations are "a stabilizing force" however (Klein and Knight 2005), and organizational norms and routines encourage maintenance of the status quo. Innovative employees may encounter barriers (e.g. conflicts with colleagues) to their new ideas when they challenge those norms (Janssen 2003).

A contextual factor that can help to overcome the costs of engaging in innovation performance is a, organizational climate favorable to risk-taking (James and McIntyre 1996). If employees perceive that a certain behavior is approved of by colleagues, their willingness to perform that particular behavior will be increased. In the case of innovation performance, it is reasonable to expect that an organizational climate that supports risk-taking will enhance the willingness of employees to engage in innovative behaviors (Ekvall 1996). Organizational members will be more likely understand that innovativeness is a desirable behavior in the organization, and will psychologically feel more secure about trial and error attempts (Yuan and Woodman 2010). It is reasonable to expect that employees that perceive a favorable risk-taking climate will enable the integration of risky behaviors, which will benefit the organizations overall innovation performance. To sum up, we propose that those organizations with a stronger risk-taking climate will show higher levels of innovation performance, compared to organizations with weaker risk-taking climates. That is,

Hypothesis 2: There is a positive and significant relationship between the risk-taking climate and innovation performance.

6.4 Method and Data

6.4.1 Sampling Frame and Data Collection

Our research hypothesis is tested on a single industry, ceramic tile manufacture, in Italy and Spain. Italian and Spanish ceramic tile producers have several things in common. Most are small and medium sized enterprises (SMEs) with a maximum of 250 workers, and generally are geographically concentrated in industrial districts (Enright and Tenti 1990). The Italian ceramic tile industrial district is located in Sassuolo (Northern Italy) and the Spanish district is in Castellón (Eastern Spain). By focusing our analysis on just one sector we can examine its particular characteristics in more depth and their influence on innovation patterns. A one sector study also reduces the range of extraneous variations in the data which could influence the constructs of interest (Coombs et al. 1996; Santarelli and Piergiovanni 1996). On the other hand, it limits generalization to other sectors but we consider that the disadvantages are outweighed by the advantages offered by this approach.

Specifically, in the production of ceramic tiles, technological accumulation is generated mainly by (1) design, construction and operation of complex production systems (scale-intensive path), and (2) knowledge, skills and techniques of chemical research emerging (science-based path). Previous studies provide evidence that Italian and Spanish ceramic tile producers are innovative (Chiva and Alegre 2009). These studies conclude that the enamel, and the tile design are the most important areas for product improvements (Hervas-Oliver et al. 2011; Meyer-Stamer et al. 2004).

The fieldwork for the present study was conducted in June to November 2004. We held surveys through personal interviews in each company. We obtained a total of 182 completed questionnaires, 101 from Spanish firms and 81 from Italian firms, which represents around 50 % of the population under study in both the Italian and the Spanish subsamples (Chamber of Commerce of Valencia 2004). The number of responses and the response rate can be considered satisfactory (Spector 1992; Williams et al. 2004). To encourage a higher response rate we offered participating firms a report of our main results.

We reduced the risk of common method variance (CMV) by collecting responses from three different respondents in each company. Collecting data from different respondents helps to control for CMV because it diminishes the effects of consistency motifs or social desirability tendencies (Podsakoff et al. 2012). Following previous research, CEOs responded to aspects of entrepreneurship (Escribá-Esteve et al.

2008); production and/or research. Production responded to questions related to innovation performance since the production manager is the person most knowledgeable about innovation activity (Calantone et al. 2002). Human resource managers responded to questions about the organizational climate (Wang and Rode 2010). Finally, to check for non-response bias, we compared sales turnover and number of employees in respondent and non-respondent firms; no significant differences were revealed.

6.4.2 Measures

Managerial risk-taking. We use the risk-taking dimension as in Covin and Slevin's (1986) entrepreneurial orientation (EO) scale. This scale was developed to reflect "the organizational processes, methods and styles that firms use to act entrepreneurially" (Lumpkin and Dess 1996, p. 139). Risk-taking is one of the three dimensions comprising the EO scale together with innovativeness and proactiveness. Specifically, risk-taking involves taking bold actions by venturing into the unknown, borrowing heavily, and/or committing significant resources to ventures in uncertain environments. Although all three dimensions are highly related, empirical evidence shows that each dimension is conceptually different and partly independent of the other dimensions (Lyon et al. 2000; Naldi et al. 2007). These items were applied using a 7-point Likert scale (see Annex).

To measure *risk-taking climate* we use the items proposed in the literature using a 7-point Likert scale (Isaksen et al. 1999) propose several items to measure employees' risk-taking climate while Amabile et al. (1996) measure how to reinforce creativity through employees' risk-taking. Our proposed scale is presented in the annex.

Innovation performance is measured using the scale provided in the OECD's (2005) Oslo Manual to assess the economic objectives of innovation. We compared innovation performance with competitors on several items (see Annex) on a 7-point Likert scale. We operationalized innovation performance as the average of three different dimensions: product innovation efficacy, process innovation efficacy, and innovation project efficiency. Product and process innovation efficacy reflects the degree of success of an innovation. Innovation project efficiency reflects the effort carried out to achieve that degree of success. These dimensions have been widely discussed in innovation research (Brown and Eisenhardt 1995; Chiesa et al. 1996).

Company size and *company location* are used as control variables. Belonging to a particular industrial district provides access to a labor market as well as a number of advantages associated with the adoption of a specific institutional framework. Therefore we control whether belonging or not to an industrial district has a significant impact on the firm's innovation performance (1 = firms located in Italy, 2 = firms located in Spain). At the same time, numerous studies suggest that firm size also affects innovation results, so we asked about the number of the firm's employees according to the four categories of firm size suggested by the European Commission (OECD 2005).

6.5 Results

Descriptive statistics are shown in Table 6.1. The average firms' innovation performance was 4.89 (S.D. = 1.07). On average, risk-taking climate was 4.84 (SD = 1.13). Managers showed a score of 3.89 (S.D. = 1.31) in the managerial risk-taking propensity scale.

The psychometric properties of the measurement scales were assessed in accordance with accepted practice (Gerbing and Anderson 1988; Tippins and Sohi 2003), including validity, reliability, and scale dimensionality. Content validity was established through a review of the literature, and interviews with four ceramic tile industry experts. We computed the coefficient alpha to assess scale reliability (Fornell and Larcker 1981). All scales achieved acceptable coefficient alphas of at least 0.70 (Table 6.1).

This study uses the paired t-test to assess whether there are differences between innovative firms and non-innovative firms when one considers different managers' and risk-taking propensities. The election of t-test analyses is due to the normality distribution of risk-taking related variables. To assess the differences between innovation performers and non-innovative performers two samples are compared: the latter corresponds to those firms that assess their "innovation performance" to be lower or equal to that of their competitors and the former corresponds to those firms evaluating their "innovation performance" as being higher in contrast to that of their competitors. The results are shown in Table 6.2.

The results are shown in Table 6.2. In Hypothesis 1, we proposed the idea that the managers' risk taking propensity would be positively related to the firms' innovation performance. Consistent with Hypothesis 1, results of the paired-samples t-test showed that there is a significant difference in the scores for innovative firms ($M = 4.13$, $S.D. = 1.25$) and non-innovative firms ($M = 2.96$; $S.D. = 1.10$); $t\ (181) = -5.22$, $p = 0.000$. These results suggest that managers' risk taking propensity does have an effect on the innovativeness of the firm. Specifically, our results suggest that when managers have a higher level of risk taking propensity, firms tend to be more innovative.

In Hypothesis 2, we proposed that the firms' risk taking climate would be positively related to the firms' innovation performance. Consistent with our hypothesis, results of the paired-samples t-test evidenced that there is a significant difference in

Table 6.1 Means, standard deviations, correlations and Cronbach's alphas of all variables

	Mean	S.D.	1	2	3	4
1. Risk-taking climate	4.84	1.13	(0.83)			
2. Managerial risk-taking propensity	3.89	1.31	0.313**	(0.74)		
3. Innovation performance	4.69	1.22	0.536**	0.479**	(0.97)	
4. Size	3.49	1.41	0.409**	0.318**	0.426**	
5. Country	1.55	0.50	−0.463**	−0.073	−0.249	−0.258**

Cronbach's alpha are shown on the diagonal. To calculate the correlation coefficients, we worked with the means of the items that make up each dimension

**Statistically significant correlation coefficient ($p < 0.01$)

Table 6.2 T-test results comparing innovative firms and non-innovative firms on managers' risk taking propensity

	Non-innovative firms			Innovative firms			95 % CI for mean difference	t	df
	M	SD	n	M	SD	n			
Manager's risk taking propensity	2.96	1.10	37	4.13	1.25	146	−1.62, −0.73	−5.22***	181
Risk taking climate	4.28	0.86	37	4.99	1.14	146	−1.10, −0.31	−3.54***	181

***p<0.001

the scores for innovative firms (M=4.99; S.D.=1.14) and non-innovative firms (M=4.28; S.D.=0.86); t (181)=−3, 54, p=0.001. These results indicate that the employees' risk taking climate has a significant effect on the firms' innovative performance. Specifically, high levels of employees' risk taking climate is related to high levels of firms' innovation performance.

6.6 Discussion

The attitude of managers towards risk-taking has received considerable attention within the literature. In part, the significance of risk-taking is due to its noteworthy effects on innovation performance. Generally, managers characterized by risk-taking behavior do not constrain their actions to the unpredictable consequences of innovation decisions. In deciding whether to allocate resources or to direct processes towards the development of new products and processes, risk-taking prone managers are more willing. This idea chimes with prior empirical studies analyzing the relationship between managerial risk-taking and innovative results (Ling et al. 2008).

However, studies anchored in the organizational climate literature have pointed out the importance of specific facets of the organizational climate to promote innovative behaviors among employees. Specifically, evidence suggests that organizations that encourage an organizational climate tolerant with risk-taking can influence employees' behaviors towards innovation, thus benefiting the organization's overall innovation performance (Gilson and Shalley 2004; Yuan and Woodman 2010). This paper takes account of this literature and ultimately shows the relationship between managers' risk-taking propensity, organizational climate, and innovation.

First, the present research provides empirical evidence that managerial risk-taking is positively related to risk-taking climate. In developing our theoretical framework we considered social cognitive and signaling theory as two theories that explain the mechanisms through which risk-taking can be transmitted from the upper to the lower echelons. While the former assumes that individuals learn vicariously, the latter assumes information asymmetry and expects managers to consciously emit signals to employees. Though based on distinct assumptions, both theories support the relevance of the manager's role in generating a climate where risk-taking is supported. Second, this study provides empirical evidence that the organizations' risk-taking climate enhances innovation performance. Scholars studying organizational climate pay attention to the distinct dimensions integrating this concept, such as innovation climate. For instance, (King et al. 2007) showed that a climate for innovation exerts a positive effect on organizational performance. However, although some studies have theoretically reasoned that a risk-taking climate can affect innovative behavior and outcomes (Ekvall 1996; Yuan and Woodman 2010), empirical tests analyzing the relationship between risk-taking climate and innovation performance are surprisingly lacking.

In sum, this study shows that both employees' risk-taking climate and managers' risk taking propensity are important determinants to explain firms' innovation performance. On the one hand, the results of this study contribute to upper echelon and other leadership behavior theories by demonstrating that managerial risk-taking is an key determinant of the firms' innovation performance. On the other hand, this study contributes to the literature on organizational climate. In this case, we show empirically that risk-taking climate is positively and significantly related to innovation performance.

This study has implications for practitioners. Risk is frequently described as an essential ingredient for the achievement of innovation. However, managers' acknowledgement of the relevance of risk taking is not sufficient to achieve organizational innovation. Managers must be able to translate their proactiveness towards risk to other employees, to encourage a creative and biased climate with the potential to generate innovative behaviors. This paper underlines the relevance of supporting risk-taking climates and their effects on innovation performance. The investigation in this paper is particularly relevant to the problem faced by many organizations in relation to manager's turnover. Organizations relying on key managers for relevant decisions are confronted by uncertainty if they leave the firm. For instance, consider a manager characterized for an affinity for decisions involving high risks. If this input is significant for innovation results then if the manager leaves the firm this would be a huge loss. It is in the interests of firms to motivate risk-taking behavior among all their employees. Lastly, in the specific context of this study, that is ceramic tile firms, it is particularly relevant that risk tolerance is widespread in companies. Most of these firms are family owned and especially vulnerable to changes of management. The manager has considerable discretion to moderate the organizational climate to recognize, assess, and tolerate innovation, risk, and creativity among employees.

This study has some limitations including the nature of the data, which were collected at one moment in time. This type of research, understood as cross-sectional, becomes problematic when data change over time. However, in future research we plan longitudinal studies to evaluate possible variations over time and solve endogeneity problems. Another limitation is that the study is focused on a single industry which means that extrapolation of results to other sectors should be done with extreme caution. We need more research on other industries. Moreover, the ceramic tile industry is characterized by SMEs, which means that managers have a large degree of discretion over innovation outcomes. Future research could focus on large enterprises where the manager's influence on innovation is usually lower and the creation of a climate of risk could have greater implications. The use of self-reported innovation performance can also be considered a limitation (Venkatraman 1989). It would be interesting to collect additional objective dependent measures to avoid possible biases and add robustness to our results. Moreover, qualitative research could also improve our research by providing a deeper understanding of the object of study (Chiva and Alegre 2009).

Lastly, it would be interesting to delve further into the black box. Decentralization of decision making has been suggested as a managerial practice that empowers employees and leaves space for novel and disruptive ideas entailing high degrees of risk (Jansen et al. 2006). Also, dynamic environments have been described as pushing firms towards the generation of innovations because of the heightened possibility of product obsolescence (Sidhu et al. 2004). Hence, further research could benefit from deeper analysis of the contingent effects of these practices in the relationship between manager's risk-taking propensity, risk-taking climate, and innovation performance.

Acknowledgments The authors are grateful for the finance received by the Spanish Ministry of Science and Innovation (ECO2011-28706, ECO2011-28749 and ECO2011-29863) to carry out this research. Moreover, acknowledgements are due to CSIC for funding Ana García-Granero and Óscar Llopis research grant (JAE-Predoc del Programa «Junta para la Ampliación de Estudios») co-financed by the ESF.

6.7 Annex

6.7.1 *Questionnaire*

Managerial risk-taking

Please rate your firm's strategic posture scale (Covin and Slevin 1989)		
Totally agree with the left column		Totally agree with the right column
1 2 3	4	5 6 7
SP1. A strong proclivity for low-risk projects (with normal and certain rates of return)	1-2-3-4-5-6-7	A strong proclivity for high-risk projects (with chances of very high returns)
In general, the top managers of my firm believe that...		
SP2. Owing to the nature of the environment, it is best to explore it gradually via timid incremental behavior	1-2-3-4-5-6-7	Owing to the nature of the environment, bold, wide-ranging acts are necessary to achieve the firm's objectives
When confronted with decision-making situations involving uncertainty, my firm...		
SP3. Typically adopts a cautious, "wait-and-see" posture in order to minimize the probability of making costly decisions	1-2-3-4-5-6-7	Typically adopts a bold, aggressive posture in order to maximize the probability of exploiting potential opportunities

Risk-taking climate

Could you please assess the importance of the following items in your organization?

Item	Literature source
ER1. Initiative often receives a favorable response here, so people feel encouraged to generate new ideas.	Isaaksen, Lauer and Ekvall (1999) and Amabile, Conti, Coon, Lazenby and Herron (1996)
ER2. People are encouraged to take risks in this organization.	
ER3. People here often venture into unknown territory.	
ER4. People here receive support and encouragement when presenting new ideas.	
ER5. Ideas that still have not been tested are usually presented.	

Innovation Performance Measurement Scale

Please state your firm performance compared to that of your competitors over the last 3 years with regard to the following items

Dimension	Item	Literature source
Product innovation effectiveness	PT1. Replacement of products being phased out	OECD (2005), Brown and Eisenhardt (1995), and Chiesa et al. (1996)
	PT2. Extension of product range within main product field through new products	
	PT3. Extension of product range outside main product field	
	PT4. Development of environment-friendly products	
	PT5. Market share evolution	
	PT6. Opening of new markets abroad	
	PT7. Opening of new domestic target groups	
Process innovation effectiveness	PS1. Improvement of production flexibility	
	PS2. Reduction of production costs by cutting labor cost per unit	
	PS3. Reduction of production costs by cutting material consumption	
	PS4. Reduction of production costs by cutting energy consumption	
	PS5. Reduction of production costs by cutting rejected production rate	
	PS6. Reduction of production costs by cutting design costs	
	PS7. Reduction of production costs by cutting production cycle	
	PS8. Improvement of product quality	
	PS9. Improvement of labor conditions	
	PS10. Reduction of environmental damage	
Project innovation efficiency	EF1. Average innovation project development time	
	EF2. Average number of innovation project working hours	
	EF3. Average cost per innovation project	
	EF4. Degree of overall satisfaction with innovation project efficiency	

References

Alexiev AS, Jansen JJ, Van den Bosch FA, Volberda HW (2010) Top management team advice seeking and exploratory innovation: the moderating role of TMT heterogeneity. J Manage Stud 47(7):1343–1364

Amabile TM, Conti R, Coon H, Lazenby J, Herron M (1996) Assessing the work environment for creativity. Acad Manage J 39(5):1154–1184

Amabile TM, Barsade SG, Mueller JS, Staw BM (2005) Affect and creativity at work. Adm Sci Q 50(3):367–403

Anderson NR, West MA (1998) Measuring climate for work group innovation: development and validation of the team climate inventory. J Organ Behav 19(3):235–258

Ashkanasy NM, Wilderom C, Peterson M (2000) Handbook of organizational culture and climate. Sage, Thousand Oaks

Bandura A (1986) Social foundations of thought and action: a social cognitive theory, 1st edn. Prentice Hall, Englewood Cliffs

Bandura A (2001) Social cognitive theory: an agentic perspective. Annu Rev Psychol 52:1–26

Bantel KA, Jackson SE (1989) Top management and innovations in banking: does the composition of the top team make a difference? Strateg Manage J 10(S1):107–124

Brown SL, Eisenhardt KM (1995) Product development: past research, present findings, and future directions. Acad Manage Rev 20(2):343–378

Calantone RJ, Cavusgil ST, Zhao Y (2002) Learning orientation, firm innovation capability, and firm performance. Ind Mark Manage 31(6):515–524

Chamber of Commerce of Valencia (2004) Informe de la nueva economía global y su incidencia en los sectores tradicionales de la Comunidad Valenciana. Chamber of Commerce of Valencia, Valencia

Chen C-J, Huang J-W (2009) Strategic human resource practices and innovation performance—the mediating role of knowledge management capacity. J Bus Res 62(1):104–114

Chiesa V, Coughlan P, Voss CA (1996) Development of a technical innovation audit. J Prod Innov Manage 13(2):105–136

Chiva R, Alegre J (2009) Investment in design and firm performance: the mediating role of design management*. J Prod Innov Manage 26(4):424–440

Cohen BD, Dean TJ (2005) Information asymmetry and investor valuation of IPOs: top management team legitimacy as a capital market signal. Strateg Manage J 26(7):683–690

Connelly BL, Certo ST, Ireland RD, Reutzel CR (2011) Signaling theory: a review and assessment. J Manage 37(1):39–67

Coombs R, Narandren P, Richards A (1996) A literature-based innovation output indicator. Res Policy 25(3):403–413

Covin JG, Slevin DP (1986) The development and testing of an organizational-level entrepreneurship scale. Front Entrepreneurship Res 1(1986):626–639

Covin JG, Slevin DP (1989) Strategic management of small firms in hostile and benign environments. Strateg Manag J 10(1):75–87

Daellenbach US, McCarthy AM, Schoenecker TS (1999) Commitment to innovation: the impact of top management team characteristics. R&D Manage 29(3):199–208

Damanpour F, Gopalakrishnan S (2001) The dynamics of the adoption of product and process innovations in organizations. J Manage Stud 38(1):45–65

Denison DR (1996) What is the difference between organizational culture and organizational climate? A native's point of view on a decade of paradigm wars. Acad Manage Rev 21(3): 619–654

Ekvall G (1996) Organizational climate for creativity and innovation. Eur J Work Organ Psychol 5(1):105–123

Elenkov DS, Judge W, Wright P (2005) Strategic leadership and executive innovation influence: an international multi-cluster comparative study. Strateg Manage J 26(7):665–682

Enright MJ, Tenti P (1990) How the diamond works: the Italian ceramic tile industry. Harv Bus Rev 68(2):90–91

Escribá-Esteve A, Sánchez-Peinado L, Sánchez-Peinado E (2008) Moderating influences on the firm's strategic orientation-performance relationship. Int Small Bus J 26(4):463–489
Fernández-Mesa A, Alegre-Vidal J, Chiva-Gómez R (2012) Orientación emprendedora, capacidad de aprendizaje organizativo y desempeño innovador. J Technol Manage Innov 7(2):157–170
Finkelstein S (1992) Power in top management teams: dimensions, measurement, and validation. Acad Manage J 35(3):505–538
Fornell C, Larcker DF (1981) Structural equation models with unobservable variables and measurement error: algebra and statistics. J Mark Res 18:382–388
Gerbing DW, Anderson JC (1988) An updated paradigm for scale development incorporating unidimensionality and its assessment. J Mark Res 25:186–192
Gilson LL, Shalley CE (2004) A little creativity goes a long way: an examination of teams' engagement in creative processes. J Manage 30(4):453–470
Grojean MW, Resick CJ, Dickson MW, Smith DB (2004) Leaders, values, and organizational climate: examining leadership strategies for establishing an organizational climate regarding ethics. J Bus Ethics 55(3):223–241
Hervas-Oliver J-L, Jackson I, Tomlinson PR (2011) 'May the ovens never grow cold': regional resilience and industrial policy in the North Staffordshire ceramics industrial district–with lessons from Sassoulo and Castellon. Policy Stud 32(4):377–395
House RJ, Shamir B (1993) Toward the integration of transformational, charismatic, and visionary theories. In: Chemers MM, Ayman R (eds) Leadership theory and research: perspectives and directions. Academic, San Diego, pp 81–107
Isaksen SG, Lauer KJ, Ekvall G (1999) Situational outlook questionnaire: a measure of the climate for creativity and change. Psychol Rep 85(2):665–674
James LR, McIntyre MD (1996) Perceptions of organizational climate. In: Individual differences and behavior in organizations. Jossey-Bass, San Francisco
Jansen JJP, Van Den Bosch FAJ, Volberda HW (2006) Exploratory innovation, exploitative innovation, and performance: effects of organizational antecedents and environmental moderators. Manage Sci 52(11):1661–1674
Janssen O (2003) Innovative behaviour and job involvement at the price of conflict and less satisfactory relations with co-workers. J Occup Organ Psychol 76(3):347–364. doi:10.1348/096317903769647210
King EB, Chermont K, West M, Dawson JF, Hebl MR (2007) How innovation can alleviate negative consequences of demanding work contexts: the influence of climate for innovation on organizational outcomes. J Occup Organ Psychol 80(4):631–645
Klein KJ, Knight AP (2005) Innovation implementation overcoming the challenge. Curr Dir Psychol Sci 14(5):243–246. doi:10.1111/j.0963-7214.2005.00373.x
Kuenzi M, Schminke M (2009) Assembling fragments into a lens: a review, critique, and proposed research agenda for the organizational work climate literature. J Manage 35(3):634–717
Ling YAN, Simsek Z, Lubatkin MH, Veiga JF (2008) Transformational leadership's role in promoting corporate entrepreneurship: examining the CEO-TMT interface. Acad Manage J 51(3):557–576
Liu K, Li J, Hesterly W, Cannella A Jr (2012) Top management team tenure and technological inventions at post-IPO biotechnology firms. J Bus Res 65(9):1349–1356
Lumpkin GT, Dess GG (1996) Clarifying the entrepreneurial orientation construct and linking it to performance. Acad Manage Rev 21(1):135. doi:10.2307/258632
Lyon DW, Lumpkin GT, Dess GG (2000) Enhancing entrepreneurial orientation research: operationalizing and measuring a key strategic decision making process. J Manage 26(5): 1055–1085
Madsen EL (2007) The significance of sustained entrepreneurial orientation on performance of firms–a longitudinal analysis. Entrep Reg Dev 19(2):185–204
March JG, Shapira Z (1987) Managerial perspectives on risk and risk taking. Manage Sci 33(11):1404–1418
Martínez-Ros E, Labeaga JM (2009) Product and process innovation: persistence and complementarities. Eur Manage Rev 6(1):64–75

Meyer-Stamer J, Maggi C, Siebel S (2004) Upgrading in the tile industry of Italy, Spain and Brazil: insights from cluster and value chain analysis. In: Schmitz H (ed) Local enterprises in the global economy: Issues of governance and upgrading. Edward Elgar Publishing, Cheltenham/Northhampton pp 174–199

Naldi L, Nordqvist M, Sjöberg K, Wiklund J (2007) Entrepreneurial orientation, risk taking, and performance in family firms. Fam Bus Rev 20(1):33–47

Naumann SE, Bennett N (2000) A case for procedural justice climate: development and test of a multilevel model. Acad Manage J 43(5):881–889

Nelson RR, Winter SG (1982) An evolutionary theory of economic change. Harvard University School Press, Cambridge, MA

OECD (2005) The measurement of scientific and technological activities Oslo Manual: guidelines for collecting and interpreting innovation data. OECD, Paris

Papadakis VM, Lioukas S, Chambers D (1998) Strategic decision-making processes: the role of management and context. Strateg Manage J 19(2):115–147

Peterson RS, Smith DB, Martorana PV, Owens PD (2003) The impact of chief executive officer personality on top management team dynamics: one mechanism by which leadership affects organizational performance. J Appl Psychol 88(5):795

Pirola-Merlo A, Mann L (2004) The relationship between individual creativity and team creativity: aggregating across people and time. J Organ Behav 25(2):235–257

Podsakoff PM, MacKenzie SB, Podsakoff NP (2012) Sources of method bias in social science research and recommendations on how to control it. Annu Rev Psychol 63:539–569

Redmond MR, Mumford MD, Teach R (1993) Putting creativity to work: effects of leader behavior on subordinate creativity. Organ Behav Hum Decis Process 55(1):120–151

Santarelli E, Piergiovanni R (1996) Analyzing literature-based innovation output indicators: the Italian experience. Res Policy 25(5):689–711

Schneider B, Reichers AE (1983) On the etiology of climates. Pers Psychol 36(1):19–39

Sidhu JS, Volberda HW, Commandeur HR (2004) Exploring exploration orientation and its determinants: some empirical evidence*. J Manage Stud 41(6):913–932

Sitkin SB, Pablo AL (1992) Reconceptualizing the determinants of risk behavior. Acad Manage Rev 17(1):9–38

Spector PE (1992) Summated rating scale construction: an introduction. Sage, Newbury Park

Spence M (1973) Job market signaling. Q J Econ 87(3):355–374

Teece DJ, Pisano G, Shuen A (1997) Dynamic capabilities and strategic management. Strateg Manage J 18(7):509–533

Tippins MJ, Sohi RS (2003) IT competency and firm performance: is organizational learning a missing link? Strateg Manage J 24(8):745–761

Vega-Jurado J, Gutiérrez-Gracia A, Fernández-de-Lucio I (2008) Analyzing the determinants of firm's absorptive capacity: beyond R&D. R&D Manage 38(4):392–405

Venkatraman N (1989) The concept of fit in strategy research: toward verbal and statistical correspondence. Acad Manage Rev 14(3):423–444

Vroom VH (1964) Work and motivation. Wiley, Oxford

Wang P, Rode JC (2010) Transformational leadership and follower creativity: The moderating effects of identification with leader and organizational climate. Hum Relat 63(8):1105–1128

Williams LJ, Gavin MB, Hartman NS (2004) Structural equation modeling methods in strategy research: applications and issues. Res Methodol Strategy Manage 1:303–346

Wu H-L (2008) When does internal governance make firms innovative? J Bus Res 61(2):141–153

Wu S, Levitas E, Priem RL (2005) CEO tenure and company invention under differing levels of technological dynamism. Acad Manage J 48(5):859–873

Yuan F, Woodman RW (2010) Innovative behavior in the workplace: the role of performance and image outcome expectations. Acad Manage J 53(2):323–342

Zahra SA, Covin JG (1995) Contextual influences on the corporate entrepreneurship-performance relationship: a longitudinal analysis. J Bus Ventur 10(1):43–58

Zhou J, George JM (2001) When job dissatisfaction leads to creativity: encouraging the expression of voice. Acad Manage J 44(4):682–696

Chapter 7
Beautiful Innovation: Understanding Management Innovation in the Spanish Arts, Heritage and Recreation Industries

Blanca de-Miguel-Molina, José-Luis Hervás-Oliver, María de-Miguel-Molina, and Rafael Boix

Abstract The objective of this paper is to gain an insight into the types of innovation that Spanish firms in arts, heritage and recreation undertook during the period 2006–2011. To achieve this aim we have examined which types of innovation – product, process, marketing and organizational–, have the highest share of the total, and how important organizational innovation is for these firms. The study is based on a sample of firms in the arts, heritage and recreation industries, based on a survey drawn up by the Spanish National Statistics Institute. Three important conclusions can be inferred from the results obtained in this paper for the Spanish case. The first is that, although these sectors are less innovative in technological product and process innovations than other industries (i.e. manufacturing and services), they have a similar share in the case of non-technological innovations (organizational and marketing). The second is that, in these sectors, organizational innovations take precedence, followed by marketing, process and product innovations. Therefore, and contrary to what is usually assumed, innovations in these industries are not focused on products. This second conclusion of the paper constitutes an important contribution to the analysis of the arts, heritage and recreation industries, where product innovation is seen as a distinctive feature. The third is that firms which carried out organizational innovations also undertook marketing innovations, showing that both types of innovations are highly correlated. This situation is not found for other types of innovations.

B. de-Miguel-Molina (✉) • J.-L. Hervás-Oliver • M. de-Miguel-Molina
Department of Business Organization, Universitat Politècnica de València, Valencia, Spain
e-mail: bdemigu@omp.upv.es; jose.hervas@omp.upv.es; mademi@omp.upv.es

R. Boix
Department of Applied Economics II – Economic Structure, Universitat de València, Valencia, Spain
e-mail: rafael.boix@uv.es

J.-L. Hervás-Oliver and M. Peris-Ortiz (eds.), *Management Innovation*,
Springer Proceedings in Business and Economics, DOI 10.1007/978-3-319-03134-7_7,
© Springer International Publishing Switzerland 2014

7.1 Introduction

Arts, heritage and recreation activities come under the umbrella of creative industries, although they have been differentiated from other creative industries in terms of their not-for-profit objectives and their aim to serve a broader social purpose (Bakhshi and Throsby 2010). Moreover, it is frequently assumed that these sectors are less innovative than other creative activities (Stam et al. 2008). The consequence of this supposition is that NACEs related to services such as arts, heritage and recreation, (codes 90, 91 and 93) are completely neglected in the Community Innovation Survey (CIS) put together by Eurostat. However, some countries have encompassed these industries, such as the Spanish National Statistics Institute, which includes "arts and recreation services" in its business innovation survey.

Additionally, it is also assumed that the majority of innovations carried out by firms in these sectors are product-centred (Kloosterman 2008). The tendency to focus on product innovation is widespread in the creative and cultural industries, and authors have given it different names such as *aesthetic*, *stylistic*, *soft* and *artistic* innovation (Alcaide-Marzal and Tortajada-Esparza 2007; Cappetta et al. 2006; Stoneman 2010; Gallenson 2008). In all cases, innovation is focused on changes in product appearance. However, in the example of museums, the few studies conducted show that they innovate in products, but also in organization and technology (Camarero et al. 2011). However, despite efforts to explain peculiarities in the creative industries, literature about innovation in the arts and cultural sector is scarce and little is known about creative-cultural service industries and their innovation patterns.

The Oslo Manual (OECD 2005) defines four types of innovations: product innovations, process innovations, marketing innovations and organisational innovations. Organizational innovation is defined as "the implementation of a new organizational method in the firm's business practices, workplace organization or external relations" (OECD 2005, p. 177), and stresses the fact that it is the result of strategic decisions taken by management (pp. 51). Similarly, marketing innovation is defined as the "implementation of a new marketing method involving significant changes in product design or packaging, product placement, product promotion or pricing" (pp. 49). This paper focuses on analyzing these types of innovation in the case of creative industries. In particular, this work explores the role of management innovation in these industries and their interplay with traditional technological innovation. The Manual defines product innovation in terms of the "introduction of a good or service that is new or significantly improved". Changes in appearance by contrast come under marketing innovations, which has created controversy in terms of creative industries.

In short, our paper's goal is to examine the nature of innovative arts and heritage sectors. To reach this objective, we have put forward two questions with reference to innovation in these industries:

RQ1: Are sectors in NACEs 90–93 less innovative than other sectors?
RQ2: Do innovations in NACEs 90–93 focus mainly on products?

Data was taken from the Innovation Survey carried out by the Spanish National Statistics Institute.

The structure used in this paper is as follows. Section 7.2 briefly summarizes the basic theory on the study of innovation in the arts, heritage and recreation industries. Section 7.3 discusses the empirical study of innovation in the NACEs 90, 91, 92 and 93, and sets out the data extracted from the innovation survey carried out by the Spanish National Statistics Institute, the variables and methodology used for the study, as well as the results obtained. Our conclusions can be found in Sect. 7.4.

7.2 Innovation in the Arts, Heritage and Recreation Industries

The arts, heritage and recreation sectors come under NACEs 90, 91, 92 and 93 (see Table 7.1). NACEs 90 and 91 are part of the Knowledge Intensive Services[1] (KIS) industries, which are those associated with the knowledge-based economy (Windrum and Tomlinson 1999; Aslesen and Isaksen 2007a; Bishop 2008; Strambach 2008). Moreover, NACEs 90, 91 and 93 are considered creative industries. The Department for Culture, Media and Sport, DCMS (2009) defined creative industries as "those industries that are based on individual creativity, skill and talent. And which have the potential to create wealth and jobs through developing intellectual property". Both the definition of creative industries as per the British Department for Culture (Pratt 2008; DCMS 2009) and the characteristics attributed to KIS sectors (Nählinder 2005; Doloreux et al. 2008; Strambach 2008; Muller and Doloreux 2009; Shearmur and Doloreux 2009) make reference to the talent and abilities of persons and firms to create knowledge (Larsen 2001; Aslesen and Isaksen 2007b). Table 7.1 shows that creative services are those related to "arts and recreation activities", thus only NACES 90, 91 and 93 can be termed as creative activities.

Innovation surveys based on the Oslo Manual (OECD 2005) are being conducted in more and more countries. However, differences in the sectors covered and measurements make benchmarking between countries difficult (Bloch and López-Bassols 2009). Eurostat draws up a survey based on the guidelines provided by the OECD, entitled the Community Innovation Survey (CIS). However, some creative services are not included in the CIS Survey. The NACE codes representing services which are not included in the Community Innovation Survey are:

- NACE 90: Creative, arts and entertainment activities.
- NACE 91: Libraries, archives, museums and other cultural activities.
- NACE 92: Gambling and betting activities.
- NACE 93: Sports activities and amusement and recreation activities.

[1] Knowledge Intensive Services are not the same as Knowledge Intensive Activities. Eurostat considers that an activity is classified as knowledge intensive if the tertiary-educated persons employed (according to ISCED97, levels 5 + 6) represent more than 33 % of the total employment in that activity.

Table 7.1 Knowledge intensive services and creative services

Service		Knowledge intensive service	Creative services
HTKIS	59 Motion picture, video and television programme production, sound recording and music publishing activities	X	X
	60 Programming and broadcasting activities	X	X
	61 Telecommunications	X	
	62 Computer programming, consultancy and related activities	X	X
	63 Information service activities	X	
	72 Scientific research and development	X	X
OKIS	50 Water transport	X	
	51 Air transport	X	
	58 Publishing activities	X	X
	64 Financial service activities, except insurance and pension funding	X	
	65 Insurance, reinsurance and pension funding, except compulsory social security	X	
	66 Activities auxiliary to financial services and insurance activities	X	
	69 Legal and accounting activities	X	
	70 Activities of head offices; management consultancy activities	X	
	71 Architectural and engineering activities; technical testing and analysis	X	X
	73 Advertising and market research	X	X
	74 Other professional, scientific and technical activities	X	X
	75 Veterinary activities	X	
	78 Employment activities	X	
	80 Security and investigation activities	X	
	84 Public administration and defence; compulsory social security	X	
	85 Education	X	
	86 Human health activities	X	
	87 Residential care activities	X	
	88 Social work activities without accommodation	X	
	90 Creative, arts and entertainment activities	X	X
	91 Libraries, archives, museums and other cultural activities	X	X
	92 Gambling and betting activities	X	
	93 Sports activities and amusement and recreation activities	X	X

Source: http://epp.eurostat.ec.europa.eu/cache/ITY_SDDS/Annexes/htec_esms_an3.pdf; De-Miguel et al. (2012), UNCTAD (2010), DCMS (2009), KEA European Affairs (2006) and Lazzeretti et al. (2008)

If the innovation statistics of countries are not included in CIS surveys, the activities of these sectors are more likely to be neglected. However, some exceptions can be found. For example, the Australian Bureau of Statistics and the Spanish National

Statistics Institute include "arts and recreation services" in their business innovation surveys. Moreover, these surveys include the four types of innovation defined in the Oslo Manual (OECD 2005): product innovations, process innovations, marketing innovations and organisational innovations. The use of these categories enables a comparison of technological and non-technological innovations. The Manual (OECD 2005) specifies the meaning of every type of innovation[2].

There is no consensus on how innovative creative industries are (Müller et al. 2009; Chapain et al. 2010; Bakhshi and McVittie 2009), although authors indicate that an important feature of creative industries is the creation of symbolic products (UNCTAD 2010). Authors have tried to contextualize innovation in the creative industries, using different descriptions like *aesthetic* (Alcaide-Marzal and Tortajada-Esparza 2007), *stylistic* (Cappetta et al. 2006) and *soft* (Stoneman 2010). In every case, innovation is focused on changes in the appearance of the product. Kloosterman (2008) confirms that, in general, innovation in cultural industries is mostly product innovation.

Stoneman (2010) labels *soft innovation* that which is "concerned with changes in products (and perhaps processes) of an aesthetic or intellectual nature, that has been ignored in the study of innovation prevalent in economics". Cappetta et al. (2006) name *stylistic innovations* those related to the fashion industry, but in this case innovations "result from the reassignment of social meaning to an existing product and/or from the change of the aesthetic characteristics of a product generating both a new product – from a physical point of view – and a new meaning". Alcaide-Marzal and Tortajada-Esparza (2007) use the term *aesthetic innovations* for fashion-oriented products (footwear), in which "appearance is the most strongly perceived value, and is its main novelty". They emphasise the importance of this kind of innovation because its result can imply that a product "can be perceived as being radically different and can displace earlier products". It is important to indicate that in the Oslo Manual (OECD 2005) changes in product appearance come under marketing innovations, although authors have pointed out that in creative industries innovations come in the product.

In their description of product innovations, Stoneman and Bakhshi (2009) distinguish between *soft* (aesthetic) and *technological* innovations. Moreover, they identify two types of soft innovation: changes in products that are aesthetic in nature (for example, new books or movies) and aesthetic innovation in goods and services that are primarily functional in nature (for example, new furniture or car models). Both types of innovation mentioned by Stoneman and Bakhshi (2009) are based on new products (new titles in books and video games, new films, new theatre productions, new advertising, new lines of clothing).

[2] "A **product innovation** is the introduction of a good or service that is new or significantly improved with respect to its characteristics or intended uses (p. 48). A **process innovation** is the implementation of a new or significantly improved production or delivery method (p. 49). A **marketing innovation** is the implementation of a new marketing method involving significant changes in product design or packaging, product placement, product promotion or pricing (p. 49). An **organisational innovation** is the implementation of a new organisational method in the firm's business practices, workplace organisation or external relations" (p. 51).

In terms of the arts and culture sector, Bakhshi and Throsby (2010) discussed the lack of studies on innovation, which has been ignored in studies conducted about creative industries. They mentioned the specific characteristics that differentiate arts and culture sectors from other creative industries: their not-for-profit nature and their aim to serve a broader social purpose. Finally, they identified four types of innovation that are common to cultural institutions in the creative arts: innovation in audience reach, in artform development, in value creation, and in business management and governance. In the case of museums, Camarero et al. (2011) established three types of innovations: technological, for example, that which is used to reach audiences, organizational and value creation. They revealed that small museums lack internal resources, such as human resources, which are necessary to engage in innovation.

As the literature on creative industries focuses on product innovations, the other types of innovations (process, organisational and marketing) are for the most part forgotten. Therefore, are innovations in arts and cultural activities mostly in products as the literature on creative and cultural industries would have us believe?

7.3 Method

7.3.1 *Sample and Variables*

The data for this study was obtained from statistics compiled by the Spanish National Statistics Institute. The population for the survey was 4,690 businesses, which included firms with more than ten employees in DIRCE 2011 (Central Business Register). The data available for NACEs 90–93 was aggregated. However, the number of businesses which made up the sample for these industries and responded to the survey was not provided.

The data was organised in a scheme that contains four groups of variables (Table 7.2), in line with the Oslo Manual framework. The four groups are the different types of innovations: product, process, organisational and marketing innovations. For each variable, values were taken for six consecutive years (2006–2011) in order to observe the degree of dynamism.

The framework for the data analysis which aimed to answer the two research questions set out in this paper is mainly descriptive, due to the lack of business microdata.

7.3.2 *Results*

In this section we answer the two research questions posed at the start of this paper:

RQ1: Are sectors in NACEs 90–93 less innovative than other sectors?
RQ2: Do innovations in NACEs 90–93 focus mainly on products?

In terms of the first research question, Fig. 7.1 shows that arts and recreation activities (NACEs 90–93) are less innovative than service and manufacturing industries with regard to technological innovations. However, the percentage of firms

Table 7.2 Variables used in the analysis of innovation types in NACEs 90–93

Variable	Measure	Number of firms in the survey (Years 2009–2011)
Product innovation	Enterprises that implemented product innovations	84
	PTI1: Enterprises that have introduced new or significantly improved products	25
	PTI2: Enterprises that have introduced new or significantly improved services	69
Process innovation	Enterprises that implemented process innovations	120
	PCI1: Enterprises that developed process innovation by improving manufacturing or production methods	54
	PCI2: Enterprises that developed process innovation by improving logistics, delivery or distribution methods	8
	PCI3: Enterprises that developed process innovation by supporting process activities	81
Organisational innovation	Enterprises that implemented organisational innovations	306
	OI1: Enterprises that introduced new methods of organising work responsibilities and decision making	187
	OI2: Enterprises that introduced new business practices for organising procedures	242
	OI3: Enterprises that introduced new methods of organising external relations	66
Marketing innovation	Enterprises that implemented marketing innovations	212
	MKI1: Enterprises that introduced significant changes to the aesthetic design or packaging	35
	MKI2: Enterprises that introduced new media or techniques for product promotion	138
	MKI3: Enterprises that introduced new methods for product placement	79
	MKI4: Enterprises that introduced new methods of pricing goods or services	98

Source: INE (Spanish National Statistics Institute). Innovation survey, available at www.ine.es

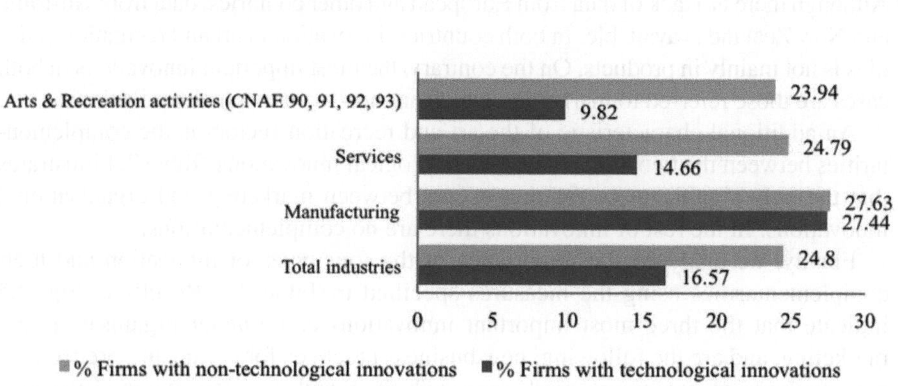

Fig. 7.1 Percentage of firms with technological and non-technological innovations (Source: INE (Spanish National Statistics Institute). Innovation survey, available at www.ine.es)

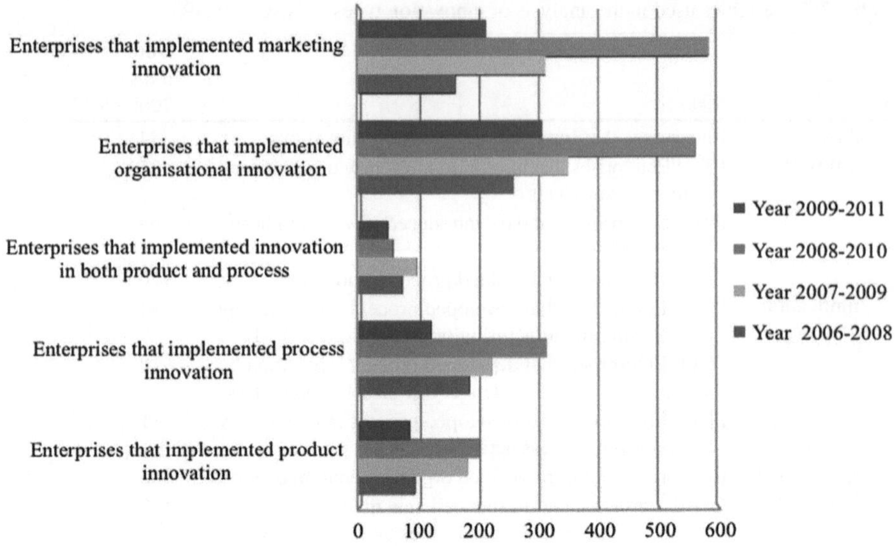

Fig. 7.2 Types of innovation in Spanish NACEs 90–93 (Source: INE (Spanish National Statistics Institute). Innovation survey, available at www.ine.es)

with non-technological innovations is similar to services and to the average of total industries. Therefore, results show that arts and recreation sectors are not less innovative (RQ1).

In order to answer the second research question, we focus on NACEs 90, 91, 92 and 93. Our aim is to verify whether innovations in these activities centre mainly on products. However, Fig. 7.2 shows that Spanish firms in arts and recreation activities innovated firstly in non-technological types of innovation (organisational and marketing), and that technological innovations were the least important. Consequently, Spanish firms in NACEs 90–93 undertook innovations that were not mostly product-based (RQ2).

Similar results can be found in other countries, as Figs. 7.3 and 7.4 demonstrate. Although there is a lack of data from European and other countries, data from Australia and New Zealand is available. In both countries, innovation in art and recreation activities is not mainly in products. On the contrary, the most important innovations in both cases are those referred to marketing, which are non-technological innovations.

An additional characteristic of the art and recreation sectors is the complementarities between the two kinds of non-technological innovations. Table 7.3 illustrates that the only significant correlation occurs between marketing and organisational innovations. In the rest of innovations there are no complementarities.

Finally, we analysed the importance of the four types of innovation and their complementarities using the measures specified in Table 7.3. Results in Fig. 7.5 indicate that the three most important innovations come under organisation and marketing, and are the following: new business practices for organising procedures, new methods of organising work responsibilities and decision making, and new

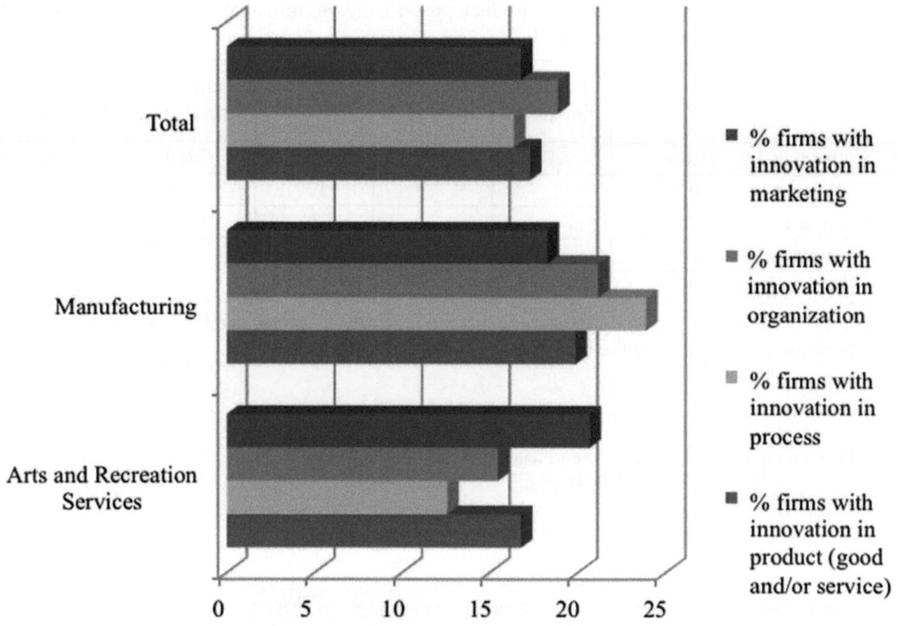

Fig. 7.3 Types of innovation in Australian NACEs 90–93, years 2010–2011 (Source: Australian Bureau of Statistics. *Innovation in Australian Businesses*)

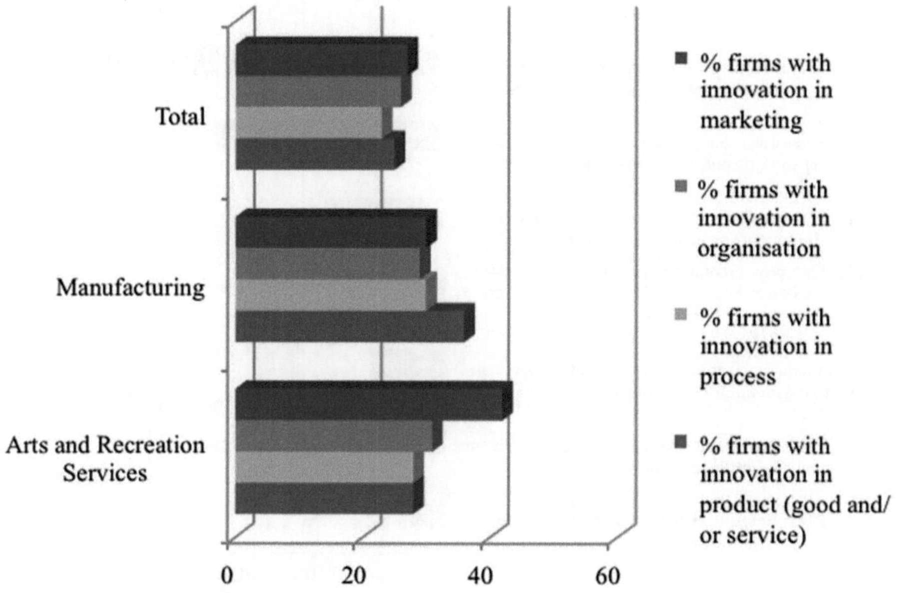

Fig. 7.4 Types of innovation in New Zealand NACEs 90–93, years 2009–2011 (Source: Statistics New Zealand. *Innovation in New Zealand: 2011*)

Table 7.3 Complementarities between product, process, organisational and marketing innovations (Pearson correlation). Years 2006–2012

	Product Innovation	Process Innovation	Product & Process Innovation	Organisational Innovation	Marketing Innovation
Product Innovation	1	.901	.373	.819	.864
Process Innovation		1	.177	.869	.897
Product & Process Innovation			1	-.221	-.137
Organisational Innovation				1	**.996****
Marketing Innovation					1

Source: INE (Spanish National Statistics Institute). Innovation survey, available at www.ine.es
**Correlation is significant at the 0.01 level (2-tailed)

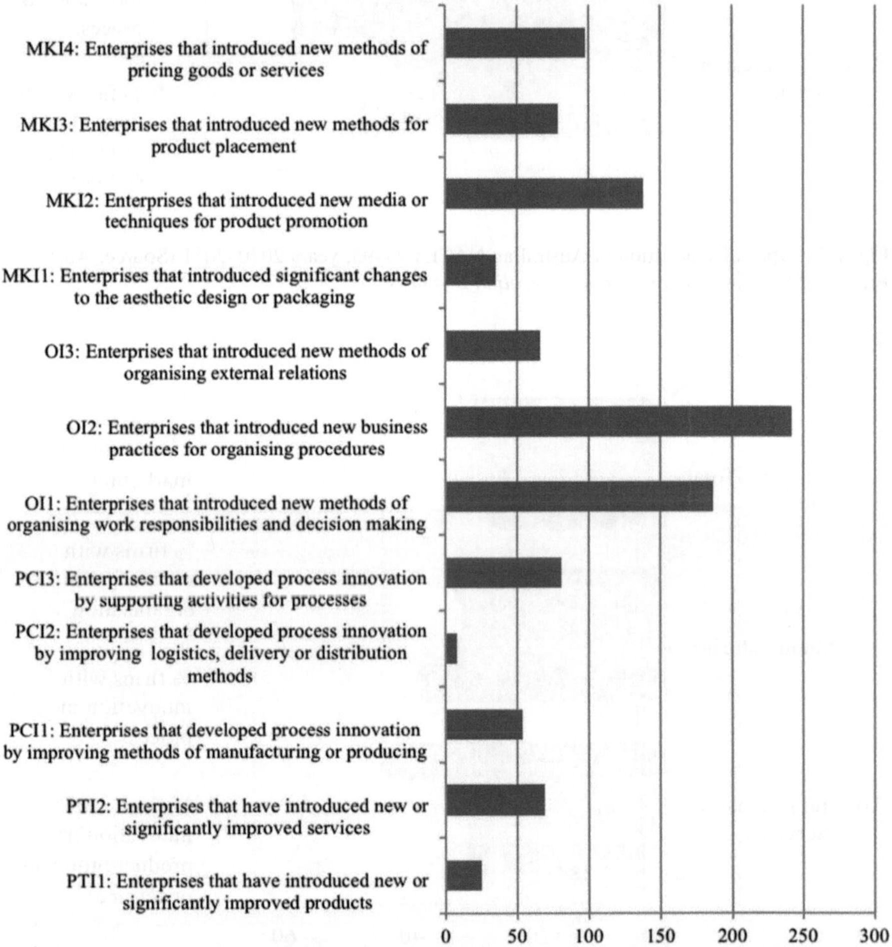

Fig. 7.5 Innovation in Spanish NACEs 90–93, year 2009–2011 (Source: INE (Spanish National Statistics Institute). Innovation survey, available at www.ine.es)

Table 7.4 Rotated component Matrix

	Component 1	Component 2
PTI1: new or significantly improved products		.942
PTI2: new or significantly improved services		.737
PCI1: process innovation by improving manufacturing or production methods	.839	
PCI2: process innovation by improving logistics, delivery or distribution methods	.794	
PCI3: process innovation by supporting process activities		.728
OI1: new methods of organising work responsibilities and decision making	.905	
OI2: new business practices for organising procedures	.896	
OI3: new methods of organising external relations		.969
MKI1: significant changes to the aesthetic design or packaging	.931	
MKI2: new media or techniques for product promotion	.911	
MKI3: new methods for product placement	.925	
MKI4: new methods of pricing goods or services	.834	

Extraction method: Principal component analysis
Rotation method: Varimax with Kaiser normalization
Source: Innovation in Spanish NACEs 90–93, years 2006–2012. Source: INE (Spanish National Statistics Institute). Innovation survey, available at www.ine.es

media or techniques for product promotion. The three least important measures are process innovation by improving logistics, delivery or distribution methods, new or significantly improved products, and significant changes to the aesthetic design or packaging. It is important to mention that the last two measures are those most closely related to innovation in creative industries.

We studied complementarities between the measures included in Table 7.2 through a factorial analysis (Table 7.4). In the analysis of the relationships between variables, two factors explained 95.67 % of the variance. The first factor alone explained 84.37 % of the variance, and the communalities' extraction data was higher than 0.9, thus all the variables reached acceptable levels of explanation. The first component showed highly positive values for marketing, organisational and process innovations, while the second component correlated the measures for product innovation positively.

7.4 Conclusions

This work focuses on analyzing the role of management innovation in creative industries. So far, no papers have addressed this topic on these specific industries. Thus, this paper's goal is novel and contributes to extend the knowledge frontier on management innovation by understanding its interplay with technological innovation in industries where less research has been done. All in all, this paper's results show that management innovation (organizational and marketing) is of utmost importance in creative industries.

Literature about arts and recreation activities indicates that these cultural-creative industries tend to be less innovative than other creative activities (Stam et al. 2008). In addition, authors have pointed out that the majority of innovations carried out by firms in these sectors focused on products (Kloosterman 2008). However, the fact that NACEs related to arts, heritage and recreation, (codes 90, 91 and 93) are neglected in the Community Innovation Survey (CIS) carried out by Eurostat, with some exceptions (like the Spanish National Statistics Institute), means that there are not enough data to make these statements. Our paper analyses these statements by studying Spanish data for arts and recreation activities.

Data for the analysis comes from the survey carried out by the Spanish National Statistics Institute (INE) on innovation activities. The framework used by the INE takes into account the different types of innovation specified in the Oslo Manual (OECD 2005), that is, product innovations, process innovations, marketing innovations and organisational innovations.

The results for NACEs 90–93 in Spain demonstrate that businesses are less innovative in technological innovations (product and process), although they are as innovative as other industries in non-technological innovations (organisation and marketing). Therefore, contrary to Stam et al. (2008), activities in art and recreation are not less innovative than other industries.

On the other hand, the results from our study confirm that innovations in marketing and organisation are more important than those in products and processes. Consequently, results do not confirm what literature indicates about innovation in creative industries being mostly in products (Kloosterman 2008; Stoneman 2010).

Acknowledgments The authors would like to thank the Spanish Ministry of Science and Innovation and the Universitat Politècnica de València (Spain) for funding this research (ECO2010-17318 MICINN Project, conducted by José-Luis Hervás-Oliver; Research Project no. 2677-UPV, conducted by Blanca de-Miguel-Molina; and PAID2012-487-UPV, conducted by Blanca de-Miguel-Molina).

References

Alcaide-Marzal J, Tortajada-Esparza E (2007) Innovation assessment in traditional industries. A proposal of aesthetic innovation indicators. Scientometrics 72(1):33–57

Aslesen HW, Isaksen A (2007a) New perspectives on knowledge-intensive services and innovation. Geografiska Annaler, Ser B: Hum Geogr 89(1):45–58

Aslesen HW, Isaksen A (2007b) Knowledge intensive business services and urban industrial development. Serv Ind J 27(3):321–338

Bakhshi H, McVittie E (2009) Creative supply-chain linkages and innovation: do the creative industries stimulate business innovation in the wider economy? Innovat Manag Pol Pract 11(2): 169–189

Bakhshi H, Throsby D (2010) Culture of innovation an economic analysis of innovation in arts and cultural organisations. NESTA, London

Bishop P (2008) Spatial spillovers and the growth of knowledge intensive services. Tijdschrift voor Economische en Sociale Geografie 99(3):281–292

Bloch C, López-Bassols V (2009) Innovation indicators. In: OECD (ed) Innovation in firms. A microeconomic perspective. OECD, Paris, pp 21–68

Camarero C, Garrido MJ, Vicente E (2011) How cultural organizations' size and funding influence innovation and performance: the case of museums. J Cultur Econ 35:247–266

Cappetta R, Cillo P, Ponti A (2006) Convergent designs in fine fashion: an evolutionary model for stylistic innovation. Res Policy 35(9):1273–1290

Chapain C, Cooke P, De Propris L, MacNeill S, Mateos-Garcia J (2010) Creative clusters and innovation. Putting creativity on the map. NESTA, London

DCMS (2009) Creative industries economic estimates statistical bulletin. Department for culture, Media & Sport, London. Available: https://www.gov.uk/government/collections/creative-industries-economic-estimates

De-Miguel-Molina B, Hervás-Oliver JL, Boix R, De-Miguel Molina M (2012) The importance of creative industry agglomerations in explaining the wealth of European regions. Eur Plann Stud 20(8):1263–1280

DIRCE (2011) Central Business Register. Spanish Statistical Office. Available: www.ine.es

Doloreux D, Amara N, Landry R (2008) Mapping regional and sectoral characteristics of knowledge-intensive business services: evidence from the province of Quebec (Canada). Growth Change 39(3):464–496

Gallenson DW (2008) Analyzing artistic innovation: the greatest breakthroughs of the twentieth century. Hist Methods: A J Quant Interdiscip Hist 41(3):111–120

KEA European Affairs (2006) The economy of culture in Europe. European Commission Directorate—General for Education and Culture, Brussels

Kloosterman RC (2008) Walls and bridges: knowledge spillover between 'superdutch' architectural firms. J Econ Geogr 8(4):545–563

Larsen JN (2001) Knowledge, human resources and social practice: the knowledge-intensive business service firm as a distributed knowledge system. Serv Ind J 21(1):81–102

Lazzeretti L, Boix R, Capone F (2008) Do creative industries cluster? Mapping creative local production systems in Italy and Spain. Ind Innov 15(5):549–567

Muller E, Doloreux D (2009) What we should know about knowledge-intensive business services. Technol Soc 31(1):64–72

Müller K, Rammer C, Trüby J (2009) The role of creative industries in industrial innovation. Innov: Manage Policy Pract 11(2):148–168

Nählinder J (2005) Innovation and employment in services. The case of knowledge business services in Sweden. Unitryck Linköping, Sweden

OECD (2005) Oslo Manual, 3rd edition. The measurement of scientific and technological activities. Proposed guidelines for collecting and interpreting technological innovation data. Available at www.oecd.org/dataoecd/35/61/2367580.pdf

Pratt AC (2008) Creative cities: the cultural industries and the creative class. Geografiska Annaler: Ser B Hum Geogr 90(2):107–117

Shearmur R, Doloreux D (2009) Place, space and distance: towards a geography of knowledge-intensive business services innovation. Ind Innov 16(1):79–102

Stam E, de Jong JPJ, Marlet G (2008) Creative industries in The Netherlands: structure, development, innovativeness and effects on urban growth. Geograkiska Annaler: Ser B Hum Geogr 90(2):119–132

Stoneman P (2010) Soft innovation. Economics, product aesthetics, and the creative industries. Oxford University Press, Oxford

Stoneman P, Bakhshi H (2009) Soft innovation. Towards a more complete picture of innovative change. NESTA, London

Strambach S (2008) Knowledge-intensive business services (KIBS) as drivers of multilevel knowledge dynamics. Int J Serv Technol Manage 10(2/3/4):152–174

UNCTAD (2010) Creative economy, report 2010. UNDP-UNCTAD, Geneva

Windrum P, Tomlinson M (1999) Knowledge-intensive services and international competitiveness: a tour country comparison. Technol Anal Strateg Manage 11(3):391–408

Chapter 8
Why Organizational Innovations Are Adopted

Angel Luís Meroño-Cerdán and Carolina López-Nicolás

Abstract Widespread agreement exists for the need to improve innovation in organizations. Despite significant steps forward, inefficiencies remain and little has been accomplished in understanding how innovation can overcome these. Current research focuses on organizational innovations. Here we focus on organizational innovations' objectives and adoption. The analysis is based on a sample of 10,796 Spanish businesses. Measures of organizational innovations and innovation objectives are based on the Oslo Manual. Statistical tests find a dynamic behavior in organizations, since 41.5 % have accomplished some organizational innovation in the period 2007–2009. Results reveal the real relation between organizational innovation objectives and adoption, with improving innovation skills being the most influential organizational innovation objective.

8.1 Introduction

Within advanced economies, production and consumption have shifted away from physical objects towards information and services, so turning the services sector into a key driver in creating competitiveness, employment and economic growth. Innovation is an important contributor to productivity and economic performance not only for manufacturers but also for service firms. In recent years, research into innovation has begun to take in services (Rubalcaba et al. 2010). Value is now created by productivity and innovation, and knowledge has become the most valuable resource (Hwang et al. 2008).

A.L. Meroño-Cerdán (✉) • C. López-Nicolás
Departamento de Organización de Empresas y Finanzas, Campus de Espinardo, Universidad de Murcia, 30100 Murcia, Spain
e-mail: angelmer@um.es; carlopez@um.es

J.-L. Hervás-Oliver and M. Peris-Ortiz (eds.), *Management Innovation*, Springer Proceedings in Business and Economics, DOI 10.1007/978-3-319-03134-7_8,

© Springer International Publishing Switzerland 2014

An organizational innovation is the implementation of a new organizational method in a firm's business practices, workplace organization or external relations (OECD 2005). Organizational innovation is a critical output for companies (Liao and Wu 2010), a source of value creation (Hwang et al. 2008) and an indicator for the intra-firm diffusion of different organizational practices (Armbruster et al. 2008). Firms may engage in innovation activity for a number of reasons: to increase performance by reducing administrative or transaction costs; to improve workplace satisfaction (and hence labor productivity); to gain access to non-tradable assets (such as non-codified external knowledge) or to reduce supply costs. Identifying enterprises' motives for innovating and their importance is helpful when examining the forces that drive innovation activity, such as competition and opportunities for entering new markets (Organization for Economic Co-operation and Development, OECD 2005).

We develop a model to understand the reasons and objectives for implementing different types of organizational innovations. The present paper seeks to predict the adoption of organizational innovations by analyzing the impact of diverse drivers. Two specific questions are addressed: What organizational innovations do organizations pursue? and What innovation objectives influence the different types of organizational innovations adopted?

The rest of this chapter is structured as follows. The salient literature on organizational innovation and the determinants of adopting innovations is reviewed. Section 8.3 describes how the data was collected from 10,796 Spanish organizations. After presenting the data analyses used, the results are discussed. The overall canonical correlation analysis provided an overview of the relationship between the goals' variables and the process innovation variables. Finally, the conclusions are summarized and the managerial implications are presented.

8.2 Literature Review

8.2.1 Organizational Innovation

According to the Oslo Manual (OECD 2005), an innovation is the implementation of a new or significantly improved product (good or service), or process, a new marketing method, or a new organizational method. This broad definition encompasses a wide range of possible innovations. The literature has argued that different types of innovation are necessary for understanding and identifying within organizations (Liao and Wu 2010). However, in practice, most innovative organizational concepts simultaneously address different types of innovations (Armsbruster et al. 2008).

An organizational innovation is the implementation of a new organizational method in the firm's business practices, workplace organization or external relations (OECD 2005). The literature states that organizational innovation is a critical output for companies (Liao and Wu 2010), a source of value creation (Hwang et al. 2008) and an indicator for the intra-firm diffusion of different organizational practices (Armbuster et al. 2008). The distinguishing features of an organizational innovation compared to other organizational changes in a firm is the implementation of an organizational method (in business practices, workplace organization or external relations) that has not been used before in the firm and is the result of strategic decisions taken by management.

Distinguishing between process and organizational innovations can be challenging since both types of innovation attempt – among other things – to decrease costs through new and more efficient concepts of production, delivery and internal organization. Many innovations therefore contain aspects of both types of innovation. For example, the introduction of new processes may also involve an initial use of new organizational methods like group working (OECD 2005). A starting point for distinguishing processes and organizational innovations is the type of activity: process innovations deal mainly with the implementation of new equipment, software and specific techniques or procedures, while organizational innovations deal primarily with people and the organization of work, whence the name of structural organizational innovations. These consist of changing responsibilities, accountability, command lines and information flows as well as the number of hierarchical levels or the divisional structure of functions.

Organizational innovation can be further differentiated through an intra-organizational and inter-organizational dimension. While intra-organizational innovations occur within an organization (such as implementation of teamwork, quality circles, continuous improvement processes or the certification of a company under ISO 9000, thus affecting departments and functions within the company), inter-organizational innovations include new organizational structures or procedures beyond a company's boundaries (Armsburster et al. 2008), like new organizational structures in an organization's environment (suppliers, customers, or competitors). Following the OECD (2005) definition of organizational innovation, the focus here is on the implementation of a new organizational method in the firm's business practices, workplace organization or external relations. Therefore, we analyze intra-organizational innovations as well as the inter-organizational dimension.

8.2.2 *Innovation Objectives*

Enterprises may engage in innovation activity for a number of reasons, which are best identified via its economic objectives (Guan et al. 2009). Their objectives may relate to products, markets, efficiency, quality or the ability to learn and to implement changes (OECD 2005). How the firm rates a number of goals that innovation (in its diverse versions) can bring within its reach relates to all its innovation activities, and should therefore be measured (Guan et al. 2009). Organizational innovations may be intended to increase a firm's performance by reducing administrative or transaction costs, to improve workplace satisfaction (and hence labor productivity), to gain access to non-tradable assets (such as non-codified external knowledge) or to reduce supply costs. Identifying enterprises' motives for innovating and their importance is helpful when examining the forces that drive innovation activity, such as competition and opportunities for entering new markets (OECD 2005).

The literature on innovation objectives suggests that different types of firms may have different primary innovation objectives (Yang and Hsiao 2009; Leiponen and Helfat 2010) due to variations in innovation patterns and the operating environment. Specifically, Guan et al. (2009) show significant differences in the importance of innovation objectives based on status (high-tech companies versus general), ownership (State-owned enterprises (SOEs) versus non- SOEs), resources for innovation (has R&D department versus no R&D department) and size (SMEs versus large enterprises). Since the literature often distinguishes between product and process innovation, Leiponen and Helfat (2010) recently observed that firms are also likely to have specific objectives for each type of innovations. For instance, process innovation objectives include goals such as reduction of labor costs, use of materials and use of energy, as well as improved manufacturing flexibility, while product innovation objectives are to replace outdated products, improve product quality, expand product assortment and enter new markets or increase market share (Leiponen and Helfat 2010). In addition, Damanpour (2010) suggests that process innovation is pursued to reduce delivery lead-time or decrease operational costs, while the objective of product innovation is to respond to customers' demand for new products or executives' desire to capture new markets.

Together with quality and cost, other innovation objectives are proposed by the extant literature. Examples include shortened response time, improved innovation skills and enhanced knowledge sharing. Kotabe and Murray (1990) suggest that both product and process innovations, as sources of long-term competitive advantage, aim to shorten innovational lead time. Companies need not only pay attention to improving efficiency and productivity, but also to develop innovation mechanisms to stimulate knowledge creation, sharing, and integration (Albers and Brewer 2003). This means that innovation activities aim to enhance knowledge sharing which is held as an innovation objective. Prior research has shown that knowledge management is an important mechanism for innovation (Lopez-Nicolas and Meroño-Cerdan 2011).

Since the maintenance, acquisition and evolution of a company's capacities depend on its innovation objectives and the resulting innovation strategy (Burgelman et al. 2001), innovation objectives may determine innovation activities and performance. For instance, Wang and Chien (2006) present a model for predicting innovation performance using technical informational resources (such as external seminar resources, external nonprofit resources, company resources and patent disclosers) and clear innovation objectives (such as improved production flexibility, reduced costs and consumption, or opening up new markets).

An understanding of the factors that drive companies to become IT innovators (Leidner et al. 2010) or innovators in general remains an important phenomenon. To date, few studies have been devoted to understanding what drives the adoption of organizational innovations.

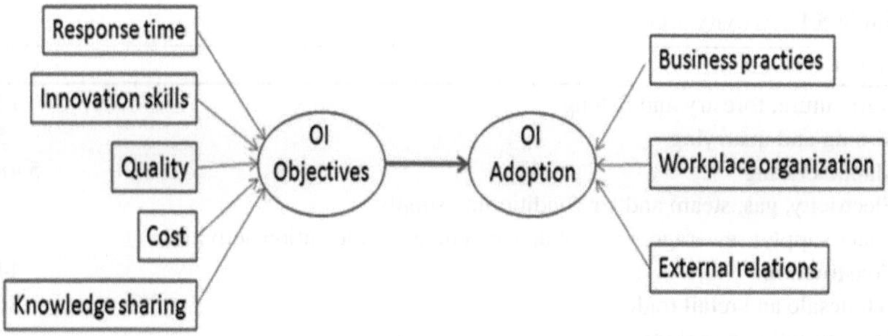

Fig. 8.1 Research model

To understand the relation between innovation objectives and innovation adoption we investigate two major questions:

1. What organizational innovation objectives do organizations pursue?
2. What innovation objectives influence the different types of organizational innovations adopted?

Figure 8.1 shows the research model:

8.3 Methodology

8.3.1 Data and Variables

Our dataset comes from the Innovation in Companies Survey for 2009 conducted by the Spanish National Statistics Institute which aims to provide information on the structure of the innovation process and to show the relations between the aforementioned process and companies' technological strategies, the factors influencing their capability to innovate and the economic performance of companies. This survey followed the methodology defined in the Oslo Manual by OECD (2005) having a mandatory nature. Questionnaires are sent via ordinary mail to a selected and representative sample of companies in terms of size and activity. Companies had a deadline of 15 days to complete the survey and each was requested to provide true information. The response rate is to 91.8 % and there is no information regarding the respondents' position in the firm. The survey inspectors are responsible for the theoretical and practical training of the personnel involved therein and for the control of the work relating to the collection of the information. An integrated information collection procedure is carried out, which consists of filtering and recording data as soon as the information is received. If required, the necessary clarifications are requested from the company. The questions address innovation activity for the period 2007–2009. A sample of 10,796 companies from different sectors (Table 8.1) and a minimum staff of 10 is extracted.

Table 8.1 Activity sector

Sector	n
Agriculture, forestry and fishing	137
Mining and quarrying	57
Manufacturing	5,488
Electricity, gas, steam and air conditioning supply	76
Water supply; sewerage, waste management and remediation activities	92
Construction	448
Wholesale and retail trade	860
Transportation and storage	240
Accommodation and food service activities	189
Information and communication	929
Financial and insurance activities	229
Real estate activities	60
Professional, scientific and technical activities	1,059
Administrative and support service activities	471
Education	54

A description of the variables included in the analysis is given below.

Organizational innovations. Adoption of organizational innovations was measured with binary variables (0 = no, 1 = yes). Three types were considered following to the Oslo Manual (OECD 2005).

- Organizational innovations in business practices involve the implementation of new methods for organizing routines and procedures for the conduct of work. These include, for example, the implementation of new practices to improve learning and knowledge sharing within the firm.
- Innovations in workplace organization involve the implementation of new methods for distributing responsibilities and decision making among employees for the division of work within and between firm activities (and organizational units), as well as new concepts for the structuring of activities, such as the integration of different business activities.
- New organizational methods in a firm's external relations involve the implementation of new ways of organizing relations with other firms or public institutions, such as the establishment of new types of collaborations with research organizations or customers, new methods of integration with suppliers, and the outsourcing or subcontracting for the first time of business activities in production, procuring, distribution, recruiting and ancillary services

Organizational innovations objectives. Items based on a Likert scale from 1 to 4 following the Oslo Manual (OECD 2005) recommendations.

- Reduce time to respond to customer or provider needs (Response time)
- Improve skills to develop new products or processes (Innovation skills)
- Increase quality of goods and services (Quality)
- Reduce unit labor costs (Cost)
- Improve information or communication sharing inside your firm or with other organizations or institutions (Knowledge sharing).

8.3.2 Analyses

This paper seeks to analyze the relations between a set of independent variables (Organizational Innovations Objectives) and a set of dependent variables (Organizational Innovations Adoption) (Fig. 8.1). Canonical analysis is a multivariate statistical technique for studying the interrelations between sets of multiple dependent or criterion variables and multiple independent or predictor variables (Hair et al. 2006). By so doing, it is likely to control for moderator effects that may exist between various dependent variables.

A between-subject multivariate analysis of variance was performed on the set of variables constituting the Objectives construct, which was the independent variable, and the Adoption construct, which was the dependent or criterion variable. The maximum number of canonical correlations (functions) between these sets of variables is the number of variables in the smaller set (Green 2000). In our study, there were five predictor variables and three criterion variables. Therefore, the number of canonical functions extracted from the analysis is three, the smallest set.

In order to gain further knowledge about the influence of objectives on each organizational innovation, a logit regression analysis was performed, including control variables (size and age) and the set of independent variables. Since the dependent variable is dichotomous, a binary logit model is developed for each organizational innovation. Logit regression tests whether coefficients are non-zero; significant and positive coefficients imply adoption facilitators, while significant and negative coefficients imply inhibitors. However, note that 'the parameters of the logit model, like those of any nonlinear regression model, are not necessarily the marginal effects we are accustomed to analyzing' (Green 2000). Actually, the marginal effect – the incremental change of the adoption probability due to unit increase of the regressor – is informed by the Odds-ratio (Exp(β)).

Goodness-of-fit is assessed in three ways. First, a likelihood ratio (LR) test, analogous to the F-test in multiple linear regressions, was conducted to examine the joint explanation power of independent variables. Second, the Hosmer – Lemershow test was used to compare the proposed model with a perfect model that can classify respondents into their respective groups correctly by comparing fitted expected values to the actual values. Third, Nagelkerke's pseudo- R2 is calculated to measure the proportion of data variation explained (Nagelkerke 1991). The logit model was also assessed in terms of the discriminating power. Based on the observation-prediction table, the rate of correct predictions by the logit model and by random guess can be computed. If the former is greater, we conclude that the logit model has a better discriminating power.

8.4 Results

During the period 2007–2009, 41.5 % of companies implemented a new organizational method (Table 8.2). The most common organizational innovation was in business practices (34.3 %) followed by workplace organization (32.9 %). Table 8.3 shows information of the distribution of companies according to the types of

Table 8.2 Organizational innovations adoption

Organizational innovation	Yes (%)
Business practices (OI1)	34.3
Workplace organization (OI2)	32.9
External Relations (OI3)	15.4
Any	41.5

Table 8.3 Combinations of organizational innovations adoption

Organizational innovation	
OI1	15.5 %
OI2	11.9 %
OI3	2.2 %
OI1 + OI3	3.0 %
OI2 + OI3	3.3 %
OI1 + OI2	35.4 %
OI1 + OI2 + OI3	28.7 %

Table 8.4 Importance of the objectives of organizational innovations adopted during 2007–2009

Objectives	Mean (1–4)
Quality	3.32
Response time	3.27
Knowledge sharing	3.11
Innovation skills	3.09
Cost	2.96

Table 8.5 Measures of overall model fit

Canonical function	Canonical correlation	Canonical R2	F-statistic
1	0.307	0.094	37.190
2	0.136	0.019	13.106
3	0.068	0.005	6.847

organizational innovations adopted. Only 29.6 % of companies adopted just one organizational innovation, and the most extended practice was a combination, especially for innovations in external relations, which rarely are used on their own.

Information related to objectives pursued by companies adopting any organizational innovation is shown in Table 8.4. The main objective is to improve quality (3.32), followed by reductions response time (3.27).

Following the guidelines suggested by Hair et al. (2006), we tested the significance of the canonical functions and the overall model fit. The overall multivariate test of significance showed that the canonical functions were statistically significant. Wilks' Lambda test is significant (F = 37.190, p = 0.000). Tables 8.5 shows the overall model fit. Three canonical functions were obtained and all were significant. The canonical R2 values support the model's validity. In the first canonical function, the independent variables explain over 9.4 % of the variance in the dependent variables. The second explains 1.9 %, and the third 0.5 %.

Table 8.6 Standard canonical coefficients and canonical loadings

Variables	Canonical loading			Canonical weight		
OI adoption	F1	F2	F3	F1	F2	F3
Business practices	0.53	0.49	−0.69	0.56	0.39	−0.73
Workplace organization	0.64	−0.76	0.08	0.63	−0.79	−0.08
External relations	0.61	0.41	0.68	0.50	0.49	0.73
OI objectives	F1	F2	F3	F1	F2	F3
Response time	0.67	−0.37	−0.32	0.26	−0.57	−0.37
Innovation skills	0.79	0.43	0.055	0.49	0.64	0.29
Quality	0.61	0.32	−0.52	0.17	0.54	−0.61
Cost	0.64	−0.50	−0.15	0.30	−0.68	0.00
Knowledge sharing	0.49	−0.08	0.68	0.28	−0.06	0.80

The relative importance of a variable in each set of variables is indicated by the canonical weight and the canonical loading extracted for the variable. The canonical weight indicates how much a variable in the predictor or criterion set contributes to the canonical function. Variables with larger weights contribute more. The canonical loading measures the simple linear correlation between an original observed variable in the predictor or criterion set and the set's linear composite and can be interpreted as a factor loading (Hair et al. 2006). Table 8.6 shows the canonical coefficients and loadings for each variable. Given the .0.3 cut-off rule (Hair et al. 2006), it is reasonable to conclude that first and second functions are more important.

The first function reveals a high correlation among all organizational innovations (Business practices, Workplace organization and External relations) and all organizational objectives, especially improvement of innovation skills. Companies keen to improve skills to develop new products or processes implement organizational innovations. The second function underlines some particularities of innovation in workplace organization. Companies adopting this innovation are more concerned about reducing costs and response time and less about innovation skills. The third function reveals differences in objectives between business practices and external relations. Improving information or communication sharing is more important in the latter, and increasing quality is the priority in the former.

Logistic regressions (Tables 8.7, 8.8 and 8.9) confirm and specify the aforementioned relationships. In each regression an initial model with control variables (size and age) only was estimated. The final model includes all variables and estimates chi-squared the increment. All the variables are significant. The significant likelihood ratio tests imply a strong relationship between dependent variables and regressors. According to the Hosmer-Lemershow test, proposed models of adoption of each organization innovation are not significantly different from a perfect one and can correctly classify observations into their respective groups (Chau and Tam 1997). Nagelkerke's pseudo-R2 reflects that about 6.0 %, 7.3 % and 5.9 % of the data variation in each organizational innovation is explained by the logit model. Finally, model overall prediction accuracy (82.6 %, 79.5 % and 63.4 % respectively) is higher than those achieved randomly (71.3 %, 67.2 % and 53.3 %).

Table 8.7 Logistic regression: business practices

	Model I			Model II		
	β	Exp(β)		β	Exp(β)	
Constant	0.632	1.882		–0.401	0.670	
Size	–0.375	0.687	***	–0.465	0.628	***
Age	0.001	1.001		0.000	1.000	
Response time				0.067	1.069	
Innovation skills				0.333	1.395	***
Quality				0.260	1.297	***
Cost				0.004	1.004	
Knowledge sharing				0.005	1.005	
Chi-squared		16.731	***		163.903	***
Δ Chi-squared					147.173	***
likelihood ratio (LR)		4119.214			3972.041	
Hosmer-Lemershow		7.730			6.815	
R^2 Nagelkerke		6 %			6 %	
% correct model		82.6 %			82.6 %	
% correct random		71.3 %			71.3 %	

$^{*}p<0.1$; $^{**}p<0.05$; $^{***}p<0.01$

Table 8.8 Logistic regression: workplace organization

	Model I			Model II		
	β	Exp(β)		β	Exp(β)	
Constant	7.198	1337.09		3.779	43.764	
Size	–0.054	0.948		–0.090	0.914	
Age	–0.003	0.997		–0.002	0.998	
Response time				0.281	1.324	**
Innovation skills				0.095	1.099	*
Quality				–0.071	0.931	
Cost				0.322	1.380	***
Knowledge sharing				0.185	1.203	***
Chi-squared		4.037			214.872	***
Δ Chi-squared					210.872	***
likelihood ratio (LR)		4563.726			4352.854	
Hosmer-Lemershow		8.863			6.345	
R^2 Nagelkerke		0.1 %			7.3 %	
% correct model		79.3 %			79.5 %	
% correct random		67.2 %			67.2 %	

$^{*}p<0.1$; $^{**}p<0.05$; $^{***}p<0.01$

Improving skills to innovate in product or process is a significant predictor in all organizational innovations. New methods in workplace organization is the function with the best fit of the three. With the exception of quality, the other innovation objectives are relevant. Innovation in workplace organization is the only organizational innovation where response time and cost are significant predictors.

Table 8.9 Logistic regression: external relations

	Model I			Model II		
	β	Exp(β)		β	Exp(β)	
Constant	−1.327	0.265		−2.768	0.063	
Size	−0.177	0.838	**	−0.235	0.790	***
Age	0.000	1.000		0.000	1.000	
Response time				0.003	1.003	
Innovation skills				0.370	1.448	***
Quality				0.092	1.097	**
Cost				0.037	1.037	
Knowledge sharing				0.219	1.245	***
Chi-squared		6.470	**		198.816	***
Δ Chi-squared					192.346	***
likelihood ratio (LR)		5905.716			5713.370	
Hosmer-Lemershow		25.832	***		0.950	
R^2 Nagelkerke		0.2 %			5.9 %	
% correct model		62.8 %			63.4 %	
% correct random		53.3 %			53.3 %	

*$p<0.1$; **$p<0.05$; ***$p<0.01$

The main difference between external relations and business practices is the importance of the role of knowledge sharing to adopt new organizational methods in external relations. Concerning the control variables, only size is a significant predictor. Smaller firms are more likely to adopt new organizational methods in business practices and external relations.

8.5 Discussion and Conclusions

This study proposes two questions concerning knowledge about situation and reasons to adopt organizational innovations. Based on recommendations of the Oslo Manual (OECD 2005), empirical analysis provides information on these issues from a representative sample of 10,796 Spanish companies. Of these, 41.5 % of them have accomplished any of the organizational innovations during the period 2007–2009.

Most common organizational innovation is in business practices (34.3 %) followed by workplace organization (32.9 %). The most common practice is to adopt a combination of them, especially in innovation in external relations, which rarely is adopted on its own. Considering this snapshot, characterized by adopting multiple organizational innovations correlations analysis, is an appropriate statistical analysis, the main benefit is the control of moderator effects that may exist among various dependent variables.

Correlation analysis confirms the influence of organizational innovation objectives on organizational innovations adoptions. Nevertheless, this research goes further and identifies more precisely the relations between innovations and objectives. Although increasing quality is generally considered most important objective of innovation, the findings reveal other objectives to be more influential. Specifically, improving innovation skills appear as the most important objective, i.e. companies especially concerned about improving skills to develop new products or processes implement organizational innovations to a higher degree. Correlation analysis also reveals some particularities of workplace organization compared to other organizational innovations. Two objectives are closely related to innovations in workplace organization: reducing response time and cost. It is precisely these objectives that can be labeled as tangible. Thus, organizational innovations in workplace organization are characterized by tangible objectives, whereas new business processes and external relations are more affected by innovation skills and quality.

This study sheds light on the actual innovation behavior by identifying the true contribution of organizational innovation objectives. Enterprises are deemed to engage in innovation activity for a number of reasons (Guan et al. 2009; OECD 2005). Extant literature on innovation objectives suggests that different types of firms may have different primary innovation objectives (Burgelman et al. 2001; Wang and Chien 2006; Yang and Hsiao 2009). The results presented here provide evidence on what innovation objectives pursue and what innovation objectives influence the different types of innovation adopted. Desires to improve innovation skills are the main objective for developing new organizational methods in business practices, workplace organization and external relations.

The main contributions of this study may be summarized as follows. First, there is a divergence between the importance given to organizational innovation objectives and their real influence on organizational innovation adoption. Improving quality is the main objective for companies adopting organizational innovations. Nevertheless, it has barely any direct influence when adopting each of the organizational innovations. Second, the most common approach is a combination of organizational innovations. Third, drivers of workplace organization innovation differ from the other organizational innovations. The latter are adopted mainly to pursue the improvement of skills to develop new products or processes. Organizational innovations play a mediator effect to increase technological innovations. The case of innovation in workplace organization seems to be definitive. Companies adopt new organizational methods in the workplace to improve operative performance by reducing time and cost.

Lastly, this study has some peculiarities derived from the data. Using an existing periodical data set from an official institution has two main advantages: first, the veracity of data (the Innovation Survey on companies is contemplated as a statistic of obligatory compliance with a posteriori controls), and, second, the opportunity to make longitudinal studies. Besides, it is a cheaper and faster way to collect data for research. In contrast, its drawback is the restriction of variables in the survey. For us, this survey fits our aim of researching drivers of organizational innovations.

The paper is exploratory in nature and can be seen as a first approach to a phenomenon scarcely analyzed by the extant literature. Thus research questions are used rather than positing hypotheses. Future research will adopt a full quantitative approach using hypotheses and including new variables. Differences in objectives of organizational innovations may arise from sectors or countries. Another important variable yet to be analyzed is performance. Thus, the main conclusions from this research may be reinforced, in particular by studying the direct and indirect benefits of organizational innovations.

References

Albers JA, Brewer S (2003) Knowledge management and the innovation process: the eco-innovation model. J Knowl Manage Pract 4(3):1–10

Armbruster H, Bikfalvi A, Kinkel S, Lay G (2008) Organizational innovation: the challenge of measuring non-technical innovation in large-scale surveys. Technovation 28(10):644–657

Burgelman R, Maidique MA, Wheelwright SC (2001) Strategic management of technology and innovation. McGraw-Hill, New York

Chau P, Tam K (1997) Factors affecting the adoption of open systems: an exploratory study. MIS Q 21(1):1–21

Damanpour F (2010) An integration of research findings of effects of firm size and market competition on product and process innovations. Br J Manage 21(4):996–1010

Green W (2000) Econometric analysis. Prentice-Hall, New York/Upper Saddle River

Guan J, Yam R, Tang E, Lau A (2009) Innovation strategy and performance during economic transition: evidences in Beijing, China. Res Policy 38(5):802–812

Hair J, Anderson R, Tatham R, Black W (2006) Multivariate data analysis, 6th edn. Prentice-Hall, New York/Upper Saddle River

Hwang HG, Chang I, Chen F, Wu S (2008) Investigation of the application of KMS for diseases classifications: a study in a Taiwanese hospital. Expert Syst Appl 34(1):725–733

Kotabe M, Murray J (1990) Linking product and process innovations and modes of international sourcing in global competition: a case of foreign multinational firms. J Int Bus Stud 21(3): 383–408

Leidner D, Preston D, Chen D (2010) An examination of the antecedents and consequences of organizational IT innovation in hospitals. J Strateg Inf Syst 19(3):154–170

Leiponen S, Helfat CE (2010) Innovation objectives, knowledge sources, and he benefits of breadth. Strateg Manage J 31(2):224–236

Liao S, Wu C (2010) System perspective of knowledge management, organizational learning, and organizational innovation. Expert Syst Appl 37(2):1096–1103

López-Nicolás C, Meroño-Cerdán AL (2011) Strategic knowledge management, innovation and performance. Int J Inf Manage 31(6):502–509

Nagelkerke N (1991) A note on a general definition of the coefficient of determination. Biometrika 78(3):691–693

OECD/Statistical Office of the European Communities (2005) Oslo manual: guidelines for collecting and interpreting innovation data, 3rd edn, The measurement of scientific and technological activities. OECD Publishing, Luxembourg

Rubalcaba L, Gallego J, Hertog P (2010) The case of market and system failures in services innovation. Serv Ind J 30(4):549–566

Wang T, Chien S (2006) Forecasting innovation performance via neural networks – a case of Taiwanese manufacturing industry. Technovation 26(5–6):635–643

Yang H, Hsiao S (2009) Mechanisms of developing innovative IT-enabled services: a case study of Taiwanese healthcare service. Technovation 29(5):327–337

Chapter 9
The Key Role of Human Resource Practices for the Promotion of Creativity and Innovation: A Spanish Case Study

Naiara Escribá-Carda, María Teresa Canet-Giner, and Francisco Balbastre-Benavent

Abstract Management literature suggests that employee creativity can contribute substantially to organisational innovation, effectiveness and survival. In addition, the ability to innovate has also emerged as a basic strategic option. Innovation models express the need to implement specific human resource practices (HRP) for the development of skills, knowledge and innovation-oriented behaviours. Human resource practices can be extremely important when organisations intend to foster creativity and innovation which are key factors in competing effectively. With this aim in mind, we suggest that the existence of certain HRP such as training, performance appraisal and reward systems have a positive effect on creativity and innovation. Our results show that there is a positive relationship between HRP and innovation in both the processes and the products in the case under study. Some HRP such as autonomy, participation, training, career plans, and organised recruitment processes, appear as being strongly linked to creativity and innovation. The most important contribution of this paper refers to the mediating effect of creativity between certain human resources practices and innovation.

9.1 Introduction

Management literature suggests that employee creativity can contribute substantially to organisational innovation, effectiveness and survival (Amabile et al. 1996; Nonaka 1991). In addition, the ability to innovate has emerged as a basic strategic option in meeting environmental changes (Baumol 2004; Danneels 2002) as firms

N. Escribá-Carda (✉) • M.T. Canet-Giner • F. Balbastre-Benavent
Universitat de València, Valencia, Spain
e-mail: Naiara.Escriba@uv.es; Teresa.Canet@uv.es; francisco.balbastre@uv.es

J.-L. Hervás-Oliver and M. Peris-Ortiz (eds.), *Management Innovation*,
Springer Proceedings in Business and Economics, DOI 10.1007/978-3-319-03134-7_9,

© Springer International Publishing Switzerland 2014

have to compete in progressively more unstable, unpredictable and uncertain environments (Andersen 2000, 2004; Brews and Hunt 1999; Bueno and Ordoñez 2004; Hendry 2000; Mintzberg and Lampel 1999; Rico 2000). In the light of this situation, our task involves identifying which factors foster creativity in order to develop innovative products and services that make firms more competitive.

Different research studies have found a positive relationship between human resource practices (HRP) and organisational performance (Becker and Gerhart 1996; Delaney and Huselid 1996; Gurbuz and Ibrahim 2011; Huselid 1995; Huselid et al. 1997; Vlachos 2008). In addition, innovation models express the need to implement tangible HRP for the development of skills, knowledge and innovation-oriented behaviours (Jackson et al. 1989; Jackson and Schuler 1988, 1995; Schuler and Jackson 1987b; Schuler and MacMillan 1984: Tang 1998). These studies lead us to the conclusion that seamless integration between HRP and a firm's general strategy is highly relevant. In particular, it is essential to obtain a real fit between HRP and innovation strategies.

According to these arguments, Beugeisdijk (2008) pointed out that the adoption of specific human resource practices has a direct effect on innovation and the global performance of the firm. Chen and Huang (2009) also established that strategic HRP are positively related to knowledge management ability as well as innovation performance. Recently, Jiang et al. (2012) analysed the relationship between HRP, employee creativity and innovation, using creativity as a mediating variable. However, the study of the mediating effect of creativity on the link between HRP and organisational innovation remains relatively unexplored, particularly in the specialist literature on management topics.

Based on the arguments set out above, two research goals can be formulated. First of all, there is a need to carry out an in-depth study into how HRP may facilitate creativity and innovation in an organisational context. The second goal centres on finding out the role played by creativity in the relationship between HRP and innovation. In other words, our objective is also to analyse to what extent creativity is necessary yet not sufficient alone for organisational innovation to take place.

To achieve our research goals and given the nature of our research topics, we used a qualitative research methodology based on the case study strategy. We carried out an in-depth analysis of a single Spanish case to explore the phenomenon, and used detailed interviews and document analysis as our main means for gathering information.

Consequently, this paper is divided into the following sections. Firstly, we review the literature which analyses creativity, innovation, and the facilitating role of HRP. As a result, some basic preliminary propositions have been put forward. Subsequently, we describe the research methods and tools underlying our research process, and analyse the results of the case after our in-depth analysis. Following this analysis, we then present the main conclusions of our research, and implications for academia and for practitioners based on our theoretical examination and empirical analysis. To conclude, the limitations of the study and possible paths for future research are stated.

9.2 Creativity, Innovation and Human Resource Practices

9.2.1 Conceptualisation and Differences Between Creativity and Innovation

Creativity and innovation have become key success factors for businesses today (Mumford et al. 2002, 2003). Yet determining the boundaries of the term 'innovation' based on literature is problematic, and the same applies to the concept or construct of 'creativity'. When reviewing literature, we found that many authors use the concept of creativity as a synonym for innovation, when in our view they are clearly differentiated terms. In our opinion, the concept of creativity is as follows: creativity has been defined by most theorists as the development of ideas about products, practices, services or procedures that are novel and potentially useful to the organisation (Amabile et al. 1996; Zhou and Shalley 2003).

Creativity can be analysed from a multilevel perspective: organisational, group and individual (Zhou and Shalley 2003). The individual perspective has been broadly analysed by literature in the field of psychology (Parjanen 2012). The literature on organisational creativity also refers to collective creative processes –the team level– that often take place in organisations when developing creative outcomes (Hargadon and Bechky 2006). Thus, teamwork becomes a key factor in order to facilitate the creative process. Organisational creativity has barely generated any interest among management researchers because it has traditionally been seen as a cognitive and individual process. However, although management research has shown little interest in creativity, the latter has been studied to a much greater extent by other disciplines. The insufficient attention paid to creativity in our field seems to be linked to the belief that it has little influence on productivity or profits.

Many of the studies about creativity have been conducted from a psychological perspective, largely in the subfields of social, environmental and organisational psychology. However, creativity has also been researched by other psychology subfields, including neurological, cognitive and experimental psychology (Borghini 2005). In our view, creativity needs to be studied in greater depth by other disciplines such as sociology, economics and management, and biology (Klijn and Tomic 2010).

Conversely, the concept of innovation has been further worked and defined by the specialist literature on management topics. As an example, Van de Ven (1986) defines innovation as a process that includes the generation, development and implementation of new ideas and behaviour. Many other authors have also defined innovation and established different categories or types of innovation. However, Martin et al. (1997) recognise that all these typologies have not been analysed equally as thoroughly and this fact may create confusion with regard to their empirical validity.

With these considerations in mind, in this paper we have adopted the definition of innovation stated in the 2005 Oslo Manual. Accordingly, innovation is defined as "the implementation of a new or significantly improved product (good or service), or process, a new marketing method, or a new organisational method in business practices, workplace organisation or external relations" (OECD-EUROSTAT, Oslo Manual 2005, p. 46).

As we have said before, and according to Shalley et al. (2004), it is important to distinguish creativity from innovation. West and Farr (1990) consider that creativity is related to the creation of the idea underlying the innovation, whereas the latter is the process encompassing the proposal and application of new ideas. In other words, creativity is the first step towards the development of innovation.

According to Mumford and Gustafson (1988), Shalley et al. (2004), and Zhou and Shalley (2003), creativity refers to the generation of new ideas which are potentially useful to the organisation but, "only when these ideas are successfully implemented at the organisation or unit level would they be considered innovation" (Shalley et al. 2004, p. 934). Therefore, this difference between the generation of novel and useful ideas and their implementation is crucial when moving from the concept of creativity to innovation. As another example, Amabile (1988) states that individual creativity and organisational innovation are clearly linked systems.

Hence creativity remains crucial for organisational innovation, though it is not enough on its own. Similarly, Martin et al. (1997) defend that innovation has a more social and applied component as its impact goes beyond the individual level, directly or indirectly affecting other individuals or the organisation as a whole. These authors also state that innovation may involve creativity, though not all innovations become creative.

In our opinion, creativity becomes an individual capability. Amabile (1988) further explores this idea by stating that creativity is the interaction of three elements, namely, technical and intellectual knowledge, creative thinking and motivation. Thus, innovation becomes the result of the application or implementation of creative ideas and, in so doing, transcends individual and group levels and turns into an organisational issue.

9.2.2 Human Resource Practices and Innovation

The adoption of a particular competitive strategy (such as differentiation through innovation) involves the use of certain HRP or, as Schuler and Jackson (1987a) put it, certain specific roles or required behaviour. This means that implementing an innovation strategy has major implications for human resource management, as this type of strategy requires people to work differently and to develop specific behaviours and skills oriented to learning (Schuler and Jackson 1987a). Other studies (Jackson and Schuler 1995; Jackson et al. 1989; Schuler and Jackson 1987a, b, 1988; Schuler and MacMillan 1984; Tang 1998) suggest that innovation models express the need for specific human resource policies and practices for the development of innovation-oriented skills, knowledge and conduct as we pointed out before.

Given the importance of innovation for companies' competitiveness, various studies have sought to identify the possible antecedents of innovation. Literature has grouped these factors as being individual, organisational and contextual, and environmental (Damanpour et al. 1989; Damanpour 1991). In terms of organisational variables, literature has stressed the role of HRP as a crucial input for organisational innovation (Jiménez and Sanz 2005; Shipton et al. 2005).

Strategic management literature has explicitly analysed the relationship between human resource management and innovation and the positive effect that a particular way of managing human resources can have on innovation (Lau and Ngo 2004; Laursen 2002; Laursen and Foss 2003; Therrien and Léonard 2003; Pini and Santangelo 2005). Innovation requires the development of active learning processes in organisations (Shipton et al. 2005; Shipton et al. 2006a, b). Some HRP, such as specific recruitment and training processes, can foster the adoption or development of specific skills oriented to continuous learning, which will lead to innovation (Perdomo et al. 2009; Tang 1998).

Therrien and Léonard (2003) developed an initial approach to the subject of our analysis (HRP and innovation). In their study, they investigated the impact of HRP as an influential factor in the development of innovation in Canadian companies. They found that HRP affected the likelihood of being innovative as well as the launch of new products. Moreover, these authors argued that companies that encourage employee participation, training and reward systems are more likely to be "first innovators". In addition, the grouping of these practices into coherent, hierarchical systems also increases the likelihood of being a relevant innovator.

Furthermore, Laursen and Foss (2003) contended that just as HRP complementarities influenced financial performance, they also had an impact on innovation performance. These authors examined the issue using an empirical study of Danish data and identified two HR systems that are innovation drivers.

Pini and Santangelo (2005) explored the impact of the use of vertical HR practices, the existence of different ways of organising research and development, and the nature of employees' skills on the likelihood of generating different types of innovation. The results of their empirical study on a sample of companies in an Italian province showed that developing innovation is a heterogeneous activity.

Authors such as Shipton et al. (2005, 2006a, b) studied the relationship between policy and practice in terms of innovation. These studies suggested that innovation is promoted and supported when companies use human resource management to buttress the various stages of the innovation process. Specifically, the longitudinal study conducted in 2005 empirically tested the effect of the introduction of sophisticated approaches on recruitment, training and assessment practices to predict product and production technology innovation outcomes, and discovered the existence of a positive relationship between HRP and product and production technology innovation. They also found that innovation is enhanced in a climate of learning and is inhibited when there is a link between appraisal and remuneration.

The results of the 2006 (a) study showed that training, socialisation, teamwork, appraisal and an exploratory approach to learning were the innovation predictors for the surveyed companies (22 manufacturing firms in the United Kingdom). Performance-based rewards combined with an exploratory learning-oriented approach were positively associated with innovation in technical systems.

In turn, the results of study by Shipton et al. (2006b) revealed that overall job satisfaction was a significant predictor of subsequent organisational innovation, even after controlling for the organisation's previous innovation and profitability.

In addition, the data suggested that the relationship between overall job satisfaction and innovation in production technology and processes was moderated by two factors: the variety of jobs and the "single status/harmonisation" commitment (these terms refer to situations where working conditions and non-monetary aspects are seen to be fair).

In 2008, Sjoerd Beugelsdijk studied the relationship between certain HRP (training and job turnover, autonomy, flexible working hours, a performance-based remuneration system, type of employment contract) and product innovation. The most relevant findings presented in this study revealed that while incremental innovation can be organised through incentive systems and training programmes, the best way of driving radical innovation is to give employees autonomy in their tasks and planning. As a result, there is a positive relationship between some HR practices and product innovation performance, using the theory of creativity as a framework. The results confirm the theoretical prediction that HR practices can be a valuable resource for companies that want to innovate.

That same year, Jiménez and Sanz (2008) empirically analysed the relationship between innovation and human resource management and the effect on a company's earnings. Their findings showed that innovation contributes positively to business performance, and that human resource management (flexible job design, autonomy, teamwork, workforce planning, performance appraisal and reward systems) boost innovation.

Other research work, such as that by Ortín and Santamaría (2009), used case studies to analyse HRP in R&D departments and examine how well they have been adapted in various companies. Data shows that some practices, such as recruitment and work organisation, have to be adapted when dealing with R&D department staff. The findings in this study support the idea that the maturity of R&D departments means that research and development activities call for better skilled and more enthusiastic workers together with the setting up of multidisciplinary networks and teams. These authors concluded that delegating HRP to R&D departments enhanced practice adaptation.

The work of Chen and Huang (2009) showed that HRP are positively related to knowledge management capacity and at the same time demonstrated that these practices have a positive effect on innovation performance. Their results provide evidence of the mediating role of knowledge management capacity between strategic HRP planning and innovation.

As part of the same line of research, De Saá-Pérez and Díaz-Díaz (2010) conducted a study to identify the related internal HR factors (the existence of a human resource plan and job stability) which might affect the innovation capacity of peripheral companies in the European Union. An empirical study of 127 Canary Island companies led to the conclusion that high commitment human resource management (promotion, appraisal for promotion, participation, a training plan, variable remuneration, job security) had a positive impact on innovation in the organisation's processes. The findings also show that the formalisation of HRP in a plan and job stability further increased process innovation.

Cheng and Mohd (2010) examined the relationship between human resource management (HRM) (performance appraisal, professional development, training, reward systems and recruitment) and the organisation's product, process and administrative innovation capacity. They noted the major role played by stimulating training in these three types of organisational innovation in Malaysian manufacturing companies.

Finally, the work of Wei et al. (2011) explored the relationship between a human resource system and product innovation. Using a contextual approach, they analysed an interactive model in which corporate culture and structure were postulated as variables which moderated the relationship between HRP and product innovation. The empirical findings for 223 Chinese companies suggest that strategic human resource management (SHRM) has a positive impact on product innovation and that this relationship is stronger in the case of companies that have a development culture in companies with a more horizontal structure.

9.2.3 *HRP and Creativity*

Research on creativity in the workplace is relatively limited. Researchers have mainly focused on the effects of contextual or organisational factors and individual differences on creativity (Zhou and Shalley 2003).

In terms of the first set of works (those that relate organisational topics and contextual factors and creativity), the research of Shalley et al. (2004) states that an employee's personal characteristics (personality and cognitive style) and some contextual factors (coordination mechanisms, goal setting and special configuration of work settings) have a positive relationship with individual creativity and that this relationship is mediated by intrinsic motivation. The works of Kalyar and Rafi (2013), Martins and Terblanche (2003), McLean (2005), and Wilson and Stokes (2005) also focused on the importance that organisational climate and culture have on an employee's creativity. A culture that favours horizontal information flows will positively affect creativity. Additionally, culture should support autonomy and teamwork. Consequently, the level of interaction among employees will increase, and the creative process will be easily developed; and as a consequence, the degree of innovation will increase.

Another factor that can have an influence on creativity is leadership style. The works of Zhang (Zhang and Bartol 2010; Zhang et al. 2012) focused on the study of how leadership styles can boost creativity, and they also analysed the role of the leader in promoting individual creativity (Cardinal 2001; Mumford et al. 2003; Sosik et al. 1998, 1999). If the leadership role becomes relevant, a leadership style oriented to empowerment will have a positive impact on individual creativity through the effect of creative process engagement and intrinsic motivation (Zhang and Bartol 2010).

In terms of organisational factors, different works focus on the degree of formalisation of the organisation and on the role that formalisation can have on creativity.

Table 9.1 Human resource practices linked to innovation and/or creativity

Practices	Relevant empirical studies
Autonomy	Shipton et al. (2006), Shalley (1991), Zhou (1998), Beugeisdijk (2008), Jiménez and Sanz (2008)
Participation	Chen and Huang (2009), Jiménez and Sanz (2005), Laursen and Foss (2003), Therrien and Léonard (2003), Zhang and Bartol (2010), Saá-Pérez and Díaz-Díaz (2010), Camelo et al. (2010)
Training	Chen and Huang (2009), Camelo et al. (2010), Perdomo et al. (2006), Therrien and Léonard (2003), Leede and Looise (2005), Lau and Ngo (2004), Shipton et al. 2005, 2006b, Staw (1995), Perdomo et al. (2006), Tang (1998), Jiménez and Sanz (2008), Saá-Pérez and Díaz-Díaz (2010), Cheng and Mohd (2010), Searle and Ball (2003)
Reward systems	Chen and Huang (2009), Jiménez and Sanz (2005), Laursen and Foss (2003), Perdomo et al. (2006), Abbey and Dickson (1983), Searle and Ball (2003), Shipton et al. (2006a), Jiang et al. (2012), Beugeisdijk (2008), Cheng and Mohd (2010)
Performance appraisal	Chen and Huang (2009), Jiménez and Sanz (2005), Laursen and Foss (2003), Therrien and Léonard (2003), Shipton et al. (2006a), George and Zhou (2001), Jiménez and Sanz (2008), Saá-Pérez and Díaz-Díaz (2010), Cheng and Mohd (2010), Camelo et al. (2010)

Traditionally, it has been assumed that formalisation would inhibit creative and innovative processes. However, some authors suggest that formalising certain procedures can help to facilitate the emergence of creative ideas (Dougherty and Tolboom 2008; Kollenscher et al. 2009; Ohly et al. 2006). In particular, the results of the work of Binyamin and Carmeli (2010) reveal that there is a positive link between the formalisation and structuring of organisational processes and creativity.

Perhaps it is the degree of formalisation of the processes' content which may have a negative influence on creativity. If processes are very strictly defined and cannot be changed, the formalisation is restrictive in terms of creativity. However, if processes are broadly formalised, which encourages the time and space in which ideas should emerge (for example, by formalising how team meetings should take place and be conducted), then formalisation can promote the development of creative processes.

Jiang et al. (2012) examined how human resource management practices relate to employee creativity and organisational innovation. Their results showed that four HRM practices (hiring and recruitment, reward, job design and teamwork) were positively related to employee creativity, while training and performance appraisal were not. Employee creativity fully mediated the relationships between those four HRM practices and organisational innovation. Results suggest that HRM practices can play an important role in managing people to promote creativity and innovation in Chinese organisations.

Table 9.1 summarises the most important literature linking HRP with creativity and HRP with innovation.

According to the literature review we suggest the following research questions and propositions:

Q1. Is creativity a mediating variable between HRP and innovation?

P1. The use of practices such as autonomy, participation, empowerment, training, the use of reward systems and performance appraisal increases the probability of generating innovation.

P2. The use of practices such as autonomy, participation, empowerment, training, the use of reward systems and performance appraisal increases or encourages creativity.

9.3 Research Methodology

We carried out a qualitative research methodology in order to answer our research question and reach our goals. Our exploratory research centred on a Valencian industrial company. The case study methodology was chosen due to the nature of the phenomenon under study (the analysis of creative processes and their interactions with organisational and contextual variables) and to the limited theoretical body that has examined the mediating role of creativity in the relationship between HRP and innovation, as this enables the researcher to develop the study in the same context where the analysed phenomenon takes place (Balbastre 2003; Pettigrew 1990; Skinner et al. 2000; Yin 1989, 1993).

Several factors were taken into account when selecting the case. Firstly, we needed a firm where innovation would become a key success factor (i.e. the company's performance mostly depends on its ability to innovate). Secondly, another selection factor was the company's attitude to HR practices, and particularly its consideration of HR practices as a basic facilitator of creativity and innovation. Thirdly, a desirable (though not necessary) condition was that the firm had clearly structured and explicit policies on human resources and innovation. Finally, ease of access to information and availability of organisational members to be interviewed were also considered as important criteria. As a result, we selected Vossloh, one of Europe's leading rail industry manufacturers, and its subsidiary in Spain (Vossloh España, S.A.) whose production plant is located in Valencia (Spain).

We used in-depth interviews as the basic technique for information gathering, and consulted other organisational documents (memorandums, organisational policies, organisational reports, etc.) to obtain evidence that could facilitate the triangulation process. Two semi-structured individual interviews were conducted with employees from different business areas. Each interview lasted approximately 70 min (see Table 9.2 for details). Both the interviews were recorded and later transcribed.

ATLAS.ti 5.0 software was applied for data analysis. Initial coding when analysing the interview transcripts focused on HRP issues and their relationship with creativity and innovation. Furthermore, in order to provide more illustrative descriptions

Table 9.2 Interviewed employees and duration

Participants	No. of interviews	Length
Project manager	1	70
Organisational development manager (HR manager)	1	70

and increase the reliability of this research we have included representative pieces of information (quotes) in the next section as examples that reveal the reasoning behind the interviewees' answers.

9.4 Case Analysis and Results

Vossloh operates globally in infrastructure and rail technology markets. The company has a flexible structure which divides its operations into two divisions: rail infrastructure, and transportation. The selected firm is one of the business units in the transportation division, and is responsible for the design, construction and maintenance of railway vehicles. From now on we will name the business unit located in Valencia only as Vossloh.

Vossloh currently produces rail fastening systems, builds new track sections and maintains the existing lines (such as the Euro-tunnel linking France and England). Vossloh's new diesel locomotives pull countless trains each day. Their products' main advantages are cost efficiency, flexibility and an attractive way of financing for both buyers and renters. In addition, Vossloh designs and manufactures passenger trains for regional and urban services and supplies electrical equipment for trams and trolleybuses. Vossloh's engineering systems and its information technology products ensure that carriers/transport companies operate economically and with maximum customer benefits.

The Vossloh Engineering Centre is committed to innovation. Top technology and optimum quality are the hallmarks of the entire range of products that are developed and produced at the Valencia plant. It is a leader in the Spanish market and exports to the USA, the UK, France, Switzerland, Israel, Algeria, Egypt, Brazil, and Yugoslavia, among other countries.

The most important survival strategy of Vossloh is based on technological innovation. Other industrial activities have been moved to emerging locations.

With regard to research and development, Vossloh is aware that R&D&I takes centre stage in this new period and cannot be the result of individual efforts. This is why Vossloh has signed cooperation agreements with universities and research centres in order to develop new designs and solutions based on the latest technologies as well as improving existing products. In the Valencia plant there are over 100 engineers involved in the research and development of products and processes. Proof of the importance of innovation for the company is the ambitious challenge of Vossloh's top management to design and build the world's first hydrogen battery-powered locomotive in Valencia.

Traditionally, the Human Resources Department did administrative tasks. However, around 10 years ago Vossloh began to implement an integrative, strategic, long-term vision. This meant that when they opted for a strategic long-term perspective, they created an integral formal plan which considered the Human Resources Department as a key player in developing their strategy. At that time, they created formal procedures linked to basic HRP (recruitment, training and motivation) and instrumental HRP (job design, workforce planning, career potential and performance appraisal, among others).

In terms of basic and instrumental policies, we must highlight the emphasis placed on recruitment processes, training, job description and potential appraisal by the HR Director.

> The recruitment process is essential; a good choice is the key. The better we identify a person's values, basic knowledge and their potential, the better the results.
>
> In order to take on people who are prepared to innovate and create we have to organise a good recruitment process, that's what my experience has taught me. Good job descriptions need to be made, a good skills analysis is also relevant and knowing where talent lies is essential. If we know where the talent or the potential talent lies we can carry out an internal recruitment process but if this is not the case, we know we need to carry out an external recruitment process. Thus, this helps us to decide what kind of recruitment process we need.
>
> We also carry out annual potential and performance appraisals to find out what their individual value to the company is. We try to analyse the potential capabilities of our employees as well.

In terms of aspects related to the relationship between HRP and innovation, and in the opinion of the HR manager, we can say that there are no HRP which are specific to each department. They are generally implemented (with specific adaptations) because of the majority of their employees' high qualifications.

Although both kinds of innovation (product and process) take place in Vossloh there is a greater impact on process innovation as this is considered vital for continuous and incremental improvement.

Training, discussion and negotiation techniques and company culture improve work processes and procedures. Product innovation is fostered by the internal and external recruitment policy, and by a proper recruitment process and retention of talent.

An in-depth analysis of the data gathered from the interviews reveals both process and product innovations. Due to the large size and maturity of the company, the firm can take advantage of greater investment opportunities to develop new, very expensive processes.

With regard to the description of HRP oriented to creativity and innovation, the HR manager recognises that they are, in general, relevant to innovation performance. Specifically, we should highlight the role of the following practices: autonomy, commitment, participation, training, a formalised and structured recruitment process, and the existence of a career plan.

Autonomy-wise, the company is structured through self-directed teams thus making commitment and autonomy the key to the development of new products and processes. There is also a high level of employee participation, as there is a formalised process in terms of inter-departmental meetings and suggestions. The R&D&I Director noted that:

> Innovation depends on several departments and formal meetings are organised where we discuss the aspects of a particular project. Sometimes, at the end of the working day we meet to discuss possible project improvements in the group.

Vossloh considers training as a key practice for the firm. Thus, the organisation assigns a great deal of resources to it:

> In order to make them creative, or turn them into leaders, workers have to be trained in specific areas, such as rail issues or negotiation techniques, in addition to basic training. This is the key to the development of talent, and long-term thinking.

Having a formalised and structured recruitment process is an important part of successful innovative companies, according to the HR manager. Innovation results lie in achieving a good recruitment process.

> Everything must be structured and formalised to leave no shadow of doubt in our managers. All aspects must be well detailed as this will ensure our employees know exactly what they have to do and how. In the case of recruitment processes, it is important to consider the job description, skills analysis and to see where talent lies as it is very important to know where the potential talent in the firm lies if we are going to an recruit internally; or if it is an external recruitment process, talent must be detected during the recruitment process.

Another variable to consider is the existence of a career plan in which employees can be continuously trained and can have real promotion options. This fact will contribute to increased happiness at work, which is essential to the promotion of creativity and innovation.

> We believe career plans are vitally important to employees. Nowadays it is very difficult to retain top talent. Thus, our company is committed to providing training in skills, behaviour, languages, etc. as a way to retain employees and avoid brain drain. We also believe that happiness at work is a prerequisite for the promotion of creativity.

On the other hand, reward systems and performance evaluation are not carried out to improve innovation performance. Therefore, we can say that these practices are not considered as relevant variables for innovation, that is, from a managerial point of view they are not seen as key factors to generate creativity and innovation.

Creativity is seen as being complementary to innovation, i.e. a prerequisite for and prior to innovation. In fact, the HR director commented:

> All our activity, not only in the HR area but throughout the organisation, aims to promote knowledge, skills, behaviours… which is what we believe will generate creativity and innovation in our engineers.

Finally, it is also relevant to highlight other organisational variables that could have an impact on creativity and innovation such as organisational culture focused on innovation and communication policy.

We observed that organisational culture is a variable which, in the opinion of the respondents, has a positive effect on innovation performance. We can affirm that in the present case there is a formalised process of employee socialisation through the transmission of an innovation culture that guides their behaviour to generate added value.

Table 9.3 Vossloh results

P1	Partially fulfilled. The consideration of variables such as employees' employability and the retention of talent are more important than incentive schemes for increasing and fostering innovation
P2	Partially fulfilled. Compensation policies are not considered relevant to the promotion of creativity
Q1	Understanding creativity as a prior condition which is complementary and innovative

> In our company, when workers join the firm they go through different phases; during the first "host" phase (3 months) we check that our recruitment process has been the best possible, then comes the training phase in areas we consider necessary. Through this process we try to convey our strategy, mission, vision and values to our employees, so they know what behaviours and attitudes are expected of them.

Another important factor observed relates to communication policy; the existence of effective communication between employees both horizontally and vertically is essential. In the present case, there are formal meetings to discuss different aspects of their jobs. Finally, another variable to consider is the extent to which HRP are structured or formalised. According to the HR Director, this is a prerequisite for the development of creativity and innovation.

Table 9.3 summarises the analysis of propositions and questions for the Vossloh case:

9.5 Conclusions and Implications

Results obtained show that some HRP such as autonomy, commitment, participation, training, formalised and structured recruitment processes, and the existence of a career plan (employability), have a clear positive impact on creativity and, as a consequence, on product and process innovation. Furthermore, in the case studied, practices such as reward systems and performance appraisal or potential appraisal, do not have an effect on creativity or innovation. We also observed that these relationships are stronger if there is an organisational culture geared towards innovation (Wei et al. 2011) and an effective communication policy (horizontally and vertically). Furthermore in the case studied, creativity acts as a mediating variable between HRP and product and process innovation.

Moreover, we also observed the existence of different variables in the case study that were not initially considered in our propositions. Some of these variables involve HRP such as formalised recruitment processes, socialisation, and the existence of a career plan (affecting the employees' commitment and happiness). Likewise, it is interesting to note that the planned, systematic and formalised nature of HRP has contributed to fostering creativity and innovation. This result coincides with the arguments put forward by Binyamin and Carmeli (2010). From this viewpoint, creativity and innovation are fostered by the firm (as opposed to being

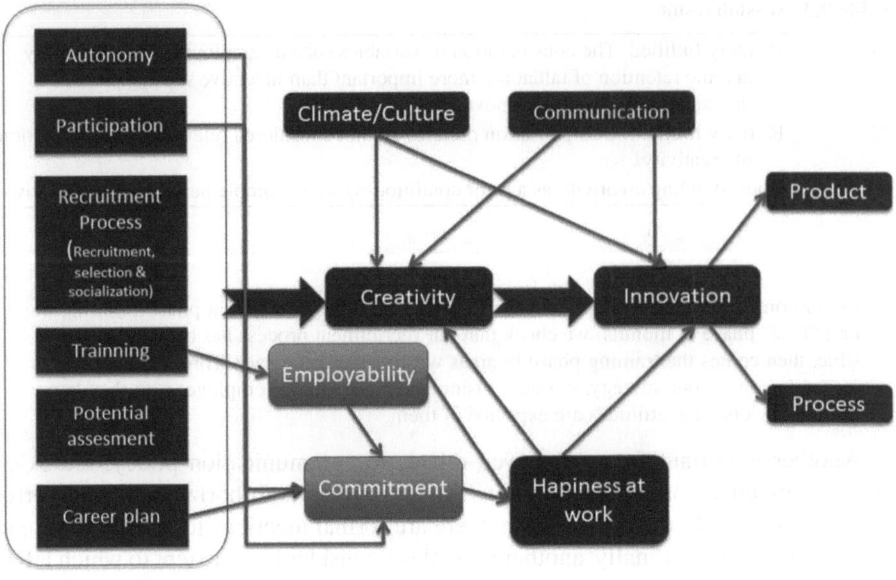

Fig. 9.1 Induced model

considered as the natural or chance outcome of these processes). Here, deliberate decision making becomes crucial to top managers. On the other hand, several contextual and organisational variables have also arisen and they should be included in our final model. These variables are the existence of a culture focused on innovation and a developed communication policy.

According to the case analysis and to the new observed variables, we developed an induced model that tries to include all the new relationships (see Fig. 9.1):

Several previous studies related certain HRP to innovation (Jiménez and Sanz 2008; Laursen and Foss 2003; Pini and Santangelo 2005; Shipton et al. 2005, 2006b). Some of them (Beugeisdijk 2008; Jimenez and Sanz 2008) propose a positive relationship between performance assessment and innovation performance. However, our research has revealed that when performance assessment is linked to reward systems there is no effect or there is even a negative effect on innovation or employee creativity. This result is supported by the specialist literature (Shipton et al. 2005).

In turn, our work revealed that creativity becomes a previous step for innovation to take place. We also observed that creativity is an individual and collective activity (teamwork) whilst innovation becomes an organisational process. As we have seen, the firm analysed understands creativity in a similar way to Amabile (1988, 1996) or West and Farr (1990). Furthermore, other recent work such as the study by Jiang et al. (2012) posted that creativity is necessary and comes before innovation, even though in our case the interviewees see links between the two concepts. This leads to the conclusion that in the case studied, creativity is considered to be a mediating variable.

The transition from individual creativity to organisational innovation needs to be further examined in future research.

To conclude with the results, we can say that our conclusions follow the trend of previous studies (Dougherty and Tolboom 2008; Kollenscher et al. 2009; Ohly et al. 2006). Specifically, different studies concluded that creativity is greater in structured conditions, such as the formalisation of the recruitment process (Chen and Huang 2009; Jiménez and Sanz 2005; Saá and Diaz 2010; among others).

This study suggests some practical implications for managers. Creativity has to be encouraged in order to generate innovation, based on autonomy, commitment, participation, training, formalised and structured recruitment processes and on the existence of a career plan (employability). These processes must be oriented to retain people with heterogeneous and flexible capabilities, and are linked to a continuous learning process. Other factors that also came up as important variables in the relationship between HRP, creativity and innovation were organisational culture and an effective communication policy. The academic implications of the paper are based on the introduction of the creativity variable as a mediating factor between HRP and innovation, an issue that needs further attention, as conclusive results have not been found, with only Jiang et al. (2012) considering HRP as a mediating variable between creativity and innovation.

This topic has scarcely been analysed in the management area. Thus, this study tries to make a contribution by identifying which variables can affect creativity and what their relationship is to innovation. In addition, after the in-depth analysis of this concept we will be closer to another future objective, which is to measure it with a quantitative scale oriented to management and not solely to psychological and individual aspects.

The main limitations concern the research methodology used. We have tried to enhance the quality of the research by drawing up and following a research protocol involving information triangulation and sending the results to the interviewees so they could verify whether our interpretations based on the information we gathered were correct or, if not, they could be amended. In addition, we enhanced scientific rigour through a review of literature that made it possible to establish a preliminary or initial theoretical framework for the object of analysis, ensuring logical consistency between the various stages of the research and thus achieving greater objectivity in the study. The choice of only one exploratory case when conducting fieldwork can be considered another limitation because its results cannot be extrapolated. However, conversely this fact permitted the researcher to analyse the phenomenon in depth and to obtain rich and comprehensive information to establish a broad framework of analysis.

More research is required to analyse how creative ideas turn into innovation performance in organisations operating in turbulent and unstable environments. Additionally, we could analyse the relationship between HRP and innovation at different levels, i.e. the effect of HRP on different working groups and innovation performance. Finally, another future line of research would be to study whether or not adapting HRP to different departments or to the different degrees of complexity of the tasks improves innovation results. Previous literature (Ortín and Santamaría

2009) suggests that HRP should be adapted to innovation departments or tasks even though the evidence obtained from our case study points in another direction. The effects of the industry and the kind of tasks carried out (product or services, degree of complexity of the tasks) could also be taken into account.

Acknowledgments Naiara Escribá Carda is a beneficiary of the programme "Research grants for Ph.D. students, under the subprogramme VLC-CAMPUS "Talent Attraction", University of Valencia.

References

Abbey A, Dickson J (1983) R&D work climate and innovation in semi-conductors. Acad Manage J 26:362–368

Amabile TM (1988) A model of creativity and innovation in organisations. In: Staw BM, Cummings LL (eds) Research in organisational behaviour, vol 10. JAI Press, Greenwich, pp 123–167

Amabile TM, Conti R, Coon H (1996) Assessing the work environment for creativity. Acad Manage J 39:292–339

Andersen TJ (2000) Strategic planning, autonomous actions and corporate performance. Long Range Plann 33:184–200

Andersen TJ (2004) Integrating decentralized strategy making and strategic planning processes in dynamic environments. J Manage Stud 41(8):1271–1299

Balbastre F (2003) La Autoevaluación Según los Modelos de Gestión de Calidad Total y el Aprendizaje en la Organización: Una Investigación de Carácter Exploratorio. Universitat de València, Valencia

Baumol W (2004) Entrepreneurial cultures and countercultures. Acad Manage Learn Educ 3(3):316–326

Becker BE, Gerhart B (1996) The impact of human resource management on organisational performance: progress and prospects. Acad Manage J 39(4):779–801

Beugeisdijk S (2008) Strategic human resource practices and product innovation. Organ Stud 29(6):821–847

Binyamin G, Carmeli A (2010) Does structuring of human resource management processes enhance employee creativity? The mediating role of psychological availability. Hum Resour Manage 49(6):999–1024

Borghini S (2005) Organisational creativity: breaking equilibrium and order to innovate. J Knowl Manage 9(4):19–33

Brews P, Hunt M (1999) Learning to plan and planning to learn: resolving the planning school/learning school debate. Strateg Manage J 20:889–913

Bueno E, Ordoñez P (2004) Innovation and learning in the knowledge-based economy: challenges for the firm. Int J Technol Manage 27(6/7):531–533

Camelo C, García J, Sousa E (2010) Facilitadores de los procesos de compartir conocimiento y su influencia sobre la innovación. Cuadernos de Economía y Dirección de la Empresa 42: 113–150

Cardinal LB (2001) Technological innovation in the pharmaceutical industry: the use of organisational control in managing research and development. Organ Sci 12(1):19–36

Chen CJ, Huang JW (2009) Strategic human resource practices and innovation performance – the mediating role of knowledge management capacity. J Bus Res 62(1):104–114

Cheng T, Mohd A (2010) Human resource management practices and organisational innovations: an empirical study in Malaysia. J Appl Bus Res 26(4):105–115

Damanpour F (1991) Organisational innovation: a meta-analysis of effects of determinants and moderators. Acad Manage J 34(3):215–240

Damanpour F, Szabat KA, Evan WM (1989) The relationship between types of innovation and organisational performance. J Manage Stud 26(6):587–601

Danneels E (2002) The dynamics of product innovation and firm competences. Strateg Manage J 23:1095–1121

De Sáa-Pérez P, Díaz-Díaz NL (2010) Human resource management and innovation in the Canary Islands: an ultra peripheral region of the European Union. Int J Hum Resour Manage 21(10):1649–1666

Delaney JT, Huselid MA (1996) The impact of human resource management practices on perceptions of organisational performance. Acad Manage J 39(4):949–969

Dougherty D, Tolboom JN (2008) Creative organising to enable organisational creativity. In: Zhou J, Shalley CE (eds) Handbook of organisational creativity. Lawrence Erlbaum Associates, Mahwah, pp 237–262

George JM, Zhou J (2001) When openness to experience and conscientiousness are related to creative behavior: an interactional approach. J Appl Psychol 86(3):513–524

Gurbuz S, Ibrahim S (2011) Impact of strategic human resource management on organisational performance: evidence from Turkey. Int J Hum Resour Manage 22(8):1803–1822

Hargadon AB, Bechky BA (2006) When collections of creatives become creative collectives: a field study of problem solving at work. Organ Sci 17:484–500

Hendry J (2000) Strategic decision making, discourse, and strategy as social practice. J Manage Stud 37(7):955–977

Huselid MA (1995) The impact of human resource management practices on turnover, productivity and corporate financial performance. Acad Manage J 38:635–672

Huselid MA, Jackson SE, Schuler RS (1997) Technical and strategic human resources management effectiveness as determinants of firm performance. Acad Manage J 40:171–188

Jackson SE, Schuler RS (1995) Understanding human resource management in the context of organisations and their environments. Annu Rev Psychol 46:237–264

Jackson SE, Schuler RS, Rivero JM (1989) Organisational characteristics as predictors of personnel practices. Pers Psychol 42:727–786

Jiang J, Wang S, Zhao S (2012) Does HRM facilitate employee creativity and organisational innovation? A study of Chinese firms. Int J Hum Resour Manage 23(19):4025–4047

Jiménez D, Sanz R (2005) Innovation and human resource management fit: an empirical study. Int J Manpow 26(4):364–398

Jiménez-Jiménez D, Sanz-Valle R (2008) Could HRM support organisational innovation? Int J Hum Resour Manage 19(7):1208–1221

Kalyar MN, Rafi N (2013) 'Organisational learning culture': an ingenious device for promoting firm's innovativeness. Serv Ind J 33(12):1135–1147

Klijn M, Tomic W (2010) A review of creativity within organisations from a psychological perspective. J Manage Dev 29(4):322–343

Kollenscher E, Ronen B, Farjoun M (2009) Architectural leadership: building a value enhancing infrastructure. Hum Syst Manage 28(1/2):35–45

Lau CM, Ngo HY (2004) The HR system, organisational culture, and product innovation. Int Bus Rev 13:685–703

Laursen K (2002) The importance of sectorial differences in the application of complementary HRM practices for innovation performance. Int J Econ Bus 9:139–156

Laursen K, Foss N (2003) New human resource management practices, complementarities and the impact on innovation performance. Camb J Econ 27:243–263

Leede (De) J, Looise JK (2005) Innovation and HRM: towards an integrated framework. Creativ Innovat Manag 14(2):108–117

Martin P, Orengo V, Martinez IM (1997) Innovación y creatividad en las organizaciones: perspectivas de análisis. Revista De Psicología Del Trabajo y Las Organizaciones 13(1):99–118

Martins EC, Terblanche F (2003) Building organisational culture that stimulates creativity and innovation. Eur J Innov Manage 6(1):64–74

McLean LD (2005) Organisational culture's influence on creativity and innovation: a review of the literature and implications for human resource development. Adv Dev Hum Resour 7(2):226–246
Mintzberg H, Lampel J (1999) Reflecting on the strategy process. Sloan Manage Rev 40(3): 21–30
Mumford M, Gustafson S (1988) Creativity syndrome: integration, application, and innovation. Psychol Bull 103(1):27–43
Mumford MD, Scott GM, Gaddis B, Strange JM (2002) Leading creative people: orchestrating expertise and relationships. Leadersh Q 13(6):705–750
Mumford MD, Connelly S, Gaddis B (2003) How creative leaders think: experimental findings and cases. Leadersh Q 14(4/5):411–432
Nonaka I (1991) The knowledge-creating company. Harv Bus Rev 69:96–104
Ohly S, Sonnentag S, Pluntke F (2006) Routinization, work characteristics and their relationship with creative and proactive behaviors. J Organ Behav 27(3):257–279
Ortín A, Santamaría L (2009) R&D managers' adaptation of firms' HRM practices. R&D Manage 39(3):271–290
OECD-EUROSTAT, Oslo Manual (2005) Guidelines for collecting and interpreting technological innovation data. OECD Publishing, Paris
Parjanen S (2012) Experiencing creativity in the organisation: from individual creativity to collective creativity. Interdiscip J Inf Knowl Manage 7:109–128
Perdomo-Ortiz J, González-Benito J, Galende J (2006) Total quality management as a Forerunner of business innovation capability. Technovation 26(10):1170–1185
Perdomo-Ortiz J, González-Benito J, Galende J (2009) An analysis of the relationship between total quality management-based human resource management practices and innovation. Int J Hum Resour Manage 20(5):1191–1218
Pettigrew AM (1990) Longitudinal field on change theory and practice. Organ Sci 1(3):267–292
Pini P, Santangelo G (2005) Innovation types and labour organisational practices: a comparison of foreign and domestic firms in the Reggio Emilia industrial districts. Econ Innov New Technol 14:251–276
Rico G (2000) Implicaciones del entorno como factor de contingencia para el diseño organizativo de las empresas multinacionales: una síntesis teórica. Revista Dirección y Organización 24:143–155
Sagiv L, Arieli S, Goldenberg J, Goldschmidt A (2009) Structure and freedom in creativity: the interplay between externally imposed structure and personal cognitive style. J Organ Behav 31(8):1086–1110
Schuler RS, Jackson SE (1987a) Linking competitive strategies with human resource management practices. Acad Manage Exec 1(3):207–219
Schuler RS, Jackson SE (1987b) Organisational strategy and organisational level as determinants of human resource practices. Hum Resour Plan 10(3):125–141
Schuler RS, Jackson SE (1988) Linking remuneration practices to innovation as a competitive strategy. Asia Pac J Hum Resour 26(2):6–20
Schuler RS, MacMillan JC (1984) Gaining competitive advantage through human resource management practices. Hum Resour Manage 23(3):241–255
Searle RH, Ball KS (2003) Supporting innovation through HR policy: evidence from the UK. Creativ Innovat Manag 12(1):50–62
Shalley CE (1991) Effects of productivity goals, creativity goals, and personal discretion on individual creativity. J Appl Psychol 76:179–185
Shalley CE, Zhou J, Oldham GR (2004) The effects of personal and contextual characteristics on creativity: where should we go from here? J Manage 30(6):933–958
Shipton HJ, Fay D, West MA, Patterson M, Birdi K (2005) Managing people to promote innovation. Creativ Innovat Manag 14(2):118–128
Shipton HJ, West MA, Dawson J, Birdi K, Patterson M (2006a) HRM as a predictor of innovation. Hum Resour Manage J 16(1):3–27
Shipton HJ, West MA, Parkes CL, Dawson JF (2006b) When promoting positive feelings pays: aggregate job satisfaction, work design features and innovation in manufacturing organisations. Eur J Work Organ Psychol 15(4):404–430

Skinner D, Tagg C, Holloway J (2000) Managers and research. The pros and cons of qualitative approaches. Manage Learn 31(2):163–179

Sosik JJ, Kahai SS, Avolio BJ (1998) Inspiring group creativity. Small Gr Res 29(1):3–31

Sosik JJ, Kahai SS, Avolio BJ (1999) Leadership style, anonymity, and creativity in group decision support systems: the mediating role of optimal flow. J Creat Behav 33(4):1–30

Staw BM (1995) Why no one really wants creativity. In: Ford CM, Gioia DA (eds) Creative action in organisations. Sage, Thousand Oaks, pp 161–166

Tang HK (1998) An integrative model of innovation in organisations. Technovation 18:297–309

Therrien P, Léonard A (2003) Empowering employees: a route to innovation. Statistics Canada, Ottawa

Van de Ven AH (1986) Central problems in the management of innovation. Manage Sci 32(5):590–607

Vlachos I (2008) The effect of human resource practices on organisational performance: evidence from Greece. Int J Hum Resour Manage 19(1):74–97

Wei L, Jun L, Herndonc NC (2011) SHRM and product innovation: testing the moderating effects of organisational culture and structure in Chinese firms. Int J Hum Resour Manage 22(1): 19–33

West MA, Farr JL (1990) Innovation at work: psychological and organisational strategies. Willey, Chichester

Wilson NC, Stokes D (2005) Managing creativity and innovation: the challenge for cultural entrepreneurs. J Small Bus Enterp Dev 12(3):366–378

Yin RK (1989) Case study research. Design and methods. Sage, Newbury Park

Yin RK (1993) Applications of case study research. Sage, Newbury Park

Zdunczyk K, Blenkinsopp J (2007) Do organisational factors support creativity and innovation in Polish firms? Eur J Innov Manage 10(1):25–40

Zhang X, Bartol KM (2010) Linking empowering leadership and employee creativity: the influence of psychological empowerment, intrinsic motivation and creative process engagement. Acad Manage J 53(1):107–128

Zhang H, Kwan HK, Zhang X, Wu LZ (2012) High core self-evaluators maintain creativity: a motivational model of abusive supervision. J Manage. doi:10.1177/0149206312460681

Zhou J (1998) Feedback valence, feedback style, task autonomy, and achievement orientation: interactive effects on creativity performance. J Appl Psychol 83:261–276

Zhou J, Shalley CE (2003) Research on employee creativity: a critical review and directions for future research. In: Martocchio J (ed) Research in personnel and human resource management. Elsevier, Oxford, UK, pp 165–217

Chapter 10
Cooperation with External Agents and Non-technological Innovations

Gloria Sánchez-González

Abstract Unlike technological innovations, non-technological ones are relatively scarcely reported in the literature. Our study analyses the influence of cooperation with different external agents: (1) suppliers, (2) customers, (3) competitors, (4) experts and consultants, and (5) universities and public research centres on the development of these types of innovation, a distinction being made between organizational and marketing innovations. The results make it clear that such cooperation is significantly favourable to the development of both, but our analysis leads us to conclude that cooperation with suppliers is the most influential of the five.

10.1 Introduction

Although research into innovation has traditionally been concentrated on advances in technology, non-technological advances have recently started to arouse growing interest. This change is due, among other things, to the impact these innovations have on firms' competitiveness, and by studying this, it is possible to identify and understand the factors governing success in technological innovations (Armbruster et al. 2006; Camisón and Villar-López 2011).

On the other hand, it is currently recognized that innovation increasingly requires the participation of a range of agents (Ritter and Gemünden 2003; Wagner and Hoegl 2006; Laursen 2011). Cooperation with other firms, institutions or other agents offers the possibility of accessing complementary resources that can contribute to a faster development of innovation activity, gain access to new markets,

G. Sánchez-González (✉)
Department of Business Administration, University of León,
Campus de Vegazana, s/n, León, Spain
e-mail: gloria.sanchez@unileon.es

J.-L. Hervás-Oliver and M. Peris-Ortiz (eds.), *Management Innovation*,
Springer Proceedings in Business and Economics, DOI 10.1007/978-3-319-03134-7_10,
© Springer International Publishing Switzerland 2014

exploit economies of scale and share costs (Ahuja 2000; Cassiman and Veugelers 2002; Hagedoorn 2002; López 2008; De Faria et al. 2010). Owing to its advantages, many studies have analysed the relationship between cooperation and technological innovations (Frenz and Ietto-Gillies 2009; Tödtling et al. 2009; De Faria et al. 2010; Tomlinson 2010; Laursen 2011). Nevertheless, it is necessary to study in greater depth the effects that such relationships may have on the development of innovations of other types, and it is interesting to ascertain which type of agent produces which effect.

In this context, the aim of the present study is to analyse the effects of cooperation with different partners in the development of non-technological innovations, with a view to thus improving our knowledge of this topic and obtaining empirical knowledge to help firms to make decisions on innovation processes. For this purpose, we propose an innovative analysis by presenting empirical evidence from Spanish firms on the effects of cooperating with (1) suppliers, (2) customers, (3) competitors, (4) experts and consultants, and (5) universities and public research centres in the development of different types of non-technological innovations, with a distinction between organizational and marketing innovations.

In order to do this, we have given the paper the following structure. Section 10.2 offers a review of the literature on non-technological innovation, again distinguishing between organizational and marketing innovations, which we relate to cooperation with different agents. Section 10.3 presents a description of the sample, the variables used for the study and the method of analysis for testing the hypotheses. Section 10.4 shows the results and their analysis. Finally, Sect. 10.5 presents the main conclusions, together with recommendations, limitations and suggestions for further research.

10.2 Review of the Literature

The importance of firms sharing knowledge and resources to develop innovations collectively has been recognized in countless studies. Such innovations are not attained only through isolated learning, but increasingly through social processes of knowledge exchange (Lundvall 1992; Edquist 1997; Edquist and Hommen 1999; Holbrook and Wolfe 2000; Landry and Amara 2001). These relationships are critical not only for access to knowledge to facilitate the diffusion of technological innovations within the firm but also because they make it possible to learn about work practices adopted by other organizations and contribute new perspectives to the firm that allow it to identify opportunities to innovate (Biemans 1991; Erickson and Jacoby 2003). It is therefore necessary for the innovation activity to have sources of information and knowledge available outside the organization, and in this situation, cooperation is a tool that firms can use to carry out their innovation activity together with external partners or agents.

On the one hand, many authors have researched the factors determining cooperation in innovation (Kleinknecht and Reijnen 1992; Fritsch and Lukas 2001; Tether 2002; Belderbos et al. 2004a), showing that the motives behind firms starting these relationships are different for each type of interlocutor. This means that they have to choose who they cooperate with very carefully (Mowery et al. 1998; Fritsch and Lukas 2001; Cassiman and Veugelers 2002; Miotti and Sachwald 2003; Belderbos et al. 2004a, b), and interest has flourished around this point (Hakanson 1993; Mowery et al. 1998; Duysters et al. 1999; Miotti and Sachwald 2003). On the other hand, it is widely recognized that cooperation favours the development of technological innovations (Laursen 2011). This entails the generation and application of new ideas to products, processes or services, offering firms opportunities to enter new markets and expand into new areas, thus obtaining competitive advantages and improving performance (Nguyen and Mothe 2008; Evangelista and Vezzani 2010).

Against this backdrop, although the literature on non-technological innovations is still somewhat lacking, they are beginning to play a more and more important role as they affect firms' competitiveness (Armbruster et al. 2008). For this reason, recent studies have encouraged research into the development of models and theories concerning non-technological innovations to extend and complete existing ones, which mainly derive from research into technological innovation (Damanpour and Aravind 2012).

Taking into account the concepts in the third edition (2005) of the Oslo Manual, non-technological innovations may be defined as those involving changes in the market behaviour (marketing innovations) or new styles of organization and management in firms (organizational innovations). Organizational innovation is a wide concept embracing the firm's structural and behavioural strategies and dimensions (Gera and Gu 2004). The concept of organizational innovation is subject to various definitions and interpretations (Lam 2005). For Black and Lynch (2004), organizational innovation includes such components as professional training, decentralization of work and flexible assignation, greater employee participation, shared rewards (such as profit sharing and share options), etc. According to Murphy (2002) and Uhlaner et al. (2007), organizational innovation involves practices of three kinds: (1) management practices (teamwork, knowledge management, flexible working arrangements, etc.), (2) production methods (changes in the organization of work: total quality management, re-engineering of business, etc.) and (3) external relations (outsourcing, networking, customer relations, etc.) These innovations can be implemented with the aim in view of improving a firm's performance, reducing administrative expenses or transaction costs, improving workers' satisfaction and therefore productivity, etc. (OECD 2005; Camisón and Villar-López 2011). Cooperation between firms promotes organizational innovations as thus they can learn more about new methods of management and organization, and so increase the number of sources to bring about innovations of this type (Yang et al. 2008). So far, however, research into the impact of cooperation on the generation of such innovations is still basically at a conceptual level, relevant empirical studies being scarce.

On the other hand, marketing innovations consist in implementing new methods of commercialization, significant changes in the design of the product or its packaging without changes to its functional characteristics, new sales channels, new promotion techniques and new pricing strategies (Deshpandé et al. 1993; Hurley and Hult 1998; OECD 2005). With these changes, firms hope to make their products or services more attractive and/or enter new markets in order to increase sales (Camisón and Villar-López 2011). Firms concentrating their attention on marketing initiatives are prone to have a greater capacity for improving customer satisfaction than their competitors (Baker and Sinkula 1999), adapting to the changing needs of the market, discovering and exploiting business ideas and accessing valuable information for the development of more competitive new products or processes (Day 1994; Rust et al. 2004). This implies the need to acquire knowledge outside the organization (Deák 2006), cooperation being very helpful for this purpose. Furthermore, cooperation on innovation can include strategic commercialization alliances to develop and later introduce new marketing concepts (OECD 2005). Innovations of this type are important because they increase the propensity to innovate and improve innovation performance (Nguyen and Mothe 2008).

In this study, it has been taken into account that firms can cooperate with different external agents, among them suppliers, customers, competitors, experts and consultants, universities and public research centres. As has already been pointed out, the aim is to ascertain the impact that these kinds of cooperation can have on the development of non-technological innovations in order to advance knowledge in this field of study.

10.2.1 Cooperation with Suppliers

Among the reasons most often mentioned in the literature for collaborating with these agents is access to wider experience and knowledge (Clark 1989; Conway 1995; Lorenzoni and Lipparini 1999; Romijn and Albaladejo 2002; Romijn and Albu 2002). Working together with suppliers can imply the need to reorganize the structure of the firm to ease the incorporation of these agents' knowledge and experience into the firm's activities and thus make it more efficient in developing such activities. To this end, in many cases the firm will have to implement new management practices to help improve the organization of work and the establishment and maintenance of relations of this type. Therefore, as a first hypothesis, it is posited that:

H1. Cooperation with suppliers favours the development of organizational innovations.

Furthermore, through this cooperation, firms may manage to reduce risks and avoid committing unnecessary mistakes in the design of innovative marketing strategies, as these agents facilitate access to complete and updated information on markets, sales channels and technologies (Fujimoto et al. 1996; Nishiguchi and Ikeda 1996; Robertson and Swan 1996). This leads to the second hypothesis:

H2. Cooperation with suppliers favours the development of marketing innovations.

10.2.2 Cooperation with Customers

This cooperation is especially valuable in the context of new technologies and/or complex products (Urban and von Hippel 1988; Neale and Corkindale 1998; Lilien et al. 2002; Tether 2002; Bogers et al. 2010), given that the customer's experience in their use may be of great help both in improving existing designs (Shaw 1994) and in coming up with new models or applications. It also brings advantages like a more directed development of the innovation process, with reduced time and costs (Jeppesen 2002), which may entail changes in the design of the internal organization to support the innovation process. This situation would imply the development of organizational innovations especially aimed at intensifying vertical and lateral communication to foment the exchange of knowledge between the firm and its customers and between the workers themselves in order to spread the ideas brought in by these external agents and apply them to the development of different types of innovation (Foss et al. 2011). The third hypothesis, then, is:

H3. Cooperation with customers favours the development of organizational innovations.

Thanks to cooperation with customers in the innovation process, it is possible to identify needs that, in many cases, the customers themselves do not yet know (Leonard and Rayport 1997). This fact allows today's firms to quickly tackle changes in consumer tastes of in modern societies (von Hippel and Katz 2002). Because of this, customers' contributions may be expected to be closely linked with the development of marketing innovations aimed at increasing sales through new strategies concerning the product, price, communication and distribution. Thus, the information provided by customers may become a key point in the design of strategies of this type. On this basis we propose the next hypothesis:

H4. Cooperation with customers favours the development of marketing innovations.

10.2.3 Cooperation with Competitors

Cooperation with competitors is frequent in high technology sectors (Mariti and Smiley 1983; Garrette and Doussauge 1995) but it can be dangerous because of the possibility of anticompetitive behaviour (Tether 2002). These risks mean that relations of this type are established in protected areas or using knowledge that is not of key importance to the firm (von Hippel 1987; Hakanson 1993). For these reasons, cooperation with competitors is mainly aimed at basic research and the setting up of standards in the sector (Gemünden et al. 1992; Tether 2002; Miotti and Sachwald 2003).

This type of cooperation is very valuable for the development of technological innovations, but in the case of organizational ones, it may not be the most appropriate practice, given that certain key knowledge of the internal organization of the firm could reach the hands of competitors in an involuntary way and jeopardize the firm's competitive position (Cassiman and Veugelers 1998; Miotti and Sachwald 2003). This leads us to the next hypothesis:

H5. Cooperation with competitors has a negative effect on the development of organizational innovations.

On the other hand, cooperation with competitors can turn out to be beneficial for the development of marketing innovations. It can be interesting, for example, to tackle the actions of firms belonging to other regions or countries and/or to develop innovative marketing campaigns for launching products or services produced jointly. Thus the following hypothesis is formed:

H6. Cooperation with competitors favours the development of marketing innovations.

10.2.4 Cooperation with Experts and Consultants

Contributions made to the firm by experts and consultants are not only concerned with saving costs but also offer the possibility of sharing experience, helping the firm to pinpoint and specify its exact needs in innovation and contribute ideas for new needs and solutions (Bessant and Rush 1995).

Furthermore, the fact of cooperating with experts outside the firm brings a different viewpoint from that of the people inside it, as often a firm's staff is very familiar with its products, processes and structures, which can put a brake on thinking up new possibilities. Experts and consultants transmit novel and different information regarding the context in which the firm and its products operate, giving rise to the generation of a greater number of innovative ideas (Bruce and Morris 1998). In turn, these novel ideas do not only contribute to the development of innovations of a technological type but may also be of great use if the firm considers changes at the organizational level or in the design of new marketing strategies.

Besides these general benefits, cooperation with these agents could prove necessary in the formulation and implementation of projects concerning improvements to structure or functioning in the organization, for which additional human resources may be needed other than those of the firm itself. In the light of this, the following hypothesis is formulated:

H7. Cooperation with experts and/or consultants favours the development of organizational innovations.

Nevertheless, the greater experience of these agents in comparison with the staff of the firm in some areas (Bessant and Rush 1995), such as aspects of trade, means that collaboration with them becomes a key factor in the design of marketing strategies allowing the firm to improve its competitive position more quickly than firms that do this alone. This argument gives rise to the hypothesis:

H8. Cooperation with experts and/or consultants favours the development of marketing innovations.

10.2.5 Cooperation with Universities and Public Research Centres

Firms may be motivated to cooperate with these agents to gain access to scientific knowledge, technical teams or new technological options (Hagedoorn 1996; Frenz and Ietto-Gillies 2009). Furthermore, unlike cooperation with other external agents, cooperation with public organizations entails no commercial risk of any kind (Cassiman and Veugelers 2002; Miotti and Sachwald 2003) as these institutions do no seek a market application for their research but are rather aimed at generating knowledge of a basic nature (Miotti and Sachwald 2003). Nevertheless, although their possible contributions may not be so valuable as those of members of the production chain (customers, suppliers, etc.), they have proved to be just as useful as those of other sources outside the firm's production system, such as competitors or consultants (Cohen et al. 2002; Fontana et al. 2003). As well as contributing additional knowledge, this kind of collaboration allows access to and recruitment of key personnel from among their teaching staff, researchers, students and graduates, necessary for the efficient development of innovation activities (Leyden and Link 1992, 1999; Burnham 1997; Meyer-Krahmer and Schmoch 1998; Link and Scott 2005; Azagra-Caro et al. 2006).

Cooperation with universities and public research centres on the development of technological innovations is positive (Meyer-Krahmer and Schmoch 1998; Beise and Stahl 1999; Link and Scott 2005; Veugelers and Cassiman 2005; Azagra-Caro et al. 2006). It is also expected for firms to approach such agents in search of new ideas, suggestions or advice concerning change in their organizational structure or in their marketing strategies, taking advantage of the knowledge that these organizations are able to contribute.

Firms decide to cooperate with universities because of the need to absorb knowledge and develop new skills or to reduce costs in infrastructure or technical staff (Jordá Borrell 2005). In this regard, the aim of the collaboration may be to analyse the firm's organizational structure and management, working on improving it and also offering the possibility of giving technical training to the firm's staff during the process. These ideas lead us to the following hypothesis:

H9. Cooperation with universities and public research centres favours the development of organizational innovations.

Moreover, as we are dealing with an external source of information, it is useful for the development of new ideas for accessing market information (Cohen et al. 2002) or to conclude existing projects on new market tactics. Likewise, as it allows firms to share risks (Cassiman and Veugelers 1998, 2005) and obtain funds for research (Bayona et al. 2000; Fontana et al. 2006), it offers the possibility of bringing down the costs of market research studies. All this forms the next hypothesis:

H10. Cooperation with universities and public research centres favours the development of marketing innovations.

10.3 Data and Methodology

In this section, we describe the sample, the variables used for the study and the econometric model.

10.3.1 Sample

This research was carried out using the data supplied by the Panel of Technological Innovation (*Panel de Innovación Tecnológica – PITEC*). This panel was created with information from Spanish firms recorded by the Technological Innovation and R&D Survey drawn up by the Spanish National Statistics Institute (*Instituto Nacional de Estadística de España – INE*). The panel supplies information permitting the analysis of the innovative behaviour of Spanish firms and its evolution. Since 2003 it has gathered data from over 7,200 firms, in both the manufacturing and service sectors.

For this research the data used correspond to the years 2006 and 2009. All the variables have been measured taking in to account the data gathered in 2009, excepting the case of the variables representing the different types of cooperation and the innovation effort, which correspond to 2006. This delay is due to the fact that the development of an innovation, regardless of its type, usually requires a more or less long period of time, so it is to be expected that the effects, both of cooperation and of the innovation effort, should be observed after a time lapse (Sánchez-González et al. 2008).

After eliminating from the sample those firms for which no complete information was available for the years in question, the final sample was made up of 10,735 firms. It included both innovating and non-innovating firms, following the recommendations of Fritsch and Lukas (2001) and Miotti and Sachwald (2003) to avoid biases in the results, such as those mentioned in other studies on the behaviour of innovating firms (Bayona et al. 2001, 2003; Tether 2002; Cassiman and Veugelers 2002; Nieto and Santamaría 2007). It should be pointed out that of the total, 2,702 firms collaborated with external agents in the development of innovations, of which 1,325 collaborated with suppliers, 937 with customers, 637 with competitors or other firms in their sector, 892 with consultants, commercial laboratories or private R&D institutes, and 1,445 with public research organizations, universities and other centres of higher education.

10.3.2 Variables

10.3.2.1 Dependent Variables

The dependent variables of the model reflect the fulfilment of two kinds of non-technological innovations: (a) organizational innovations and (b) marketing innovations. The former were measured by means of a dichotomous variable taking the value of 1 if the firm claims to have obtained any innovation of this type in the period 2007–2009, and 0 if not. Firms may have obtained organizational innovations if they developed new business practices in the organization of work or in company procedure, new methods of workplace organization, better sharing responsibilities and decision making, and/or new methods of managing external relations with other firms or public institutions.

For the case of marketing innovations, we also used a dichotomous variable, with the value 1 if the firm claimed to have carried out some kind of marketing innovation during the period 2007–2009 or 0 otherwise. These innovations may be due to significant changes in the design or packaging of a product, new techniques or promotion channels, new methods for finding their market niches or new sales channels and/or new ways of fixing prices for goods or services.

10.3.2.2 Explicative Variables

The explicative variables of the model used were cooperation with five different types of external agents: (a) suppliers, (b) customers, (c) competitors -or other firms in the sector-, (d) experts or consultants -commercial laboratories or private R&D institutes- and (e) universities and other centres of higher education, and public research centres. These relationships were measured using dichotomous variables with the value of 1 if the firm claimed to have cooperated with any of these agents during the period 2004–2006, or 0 if not (Miotti and Sachwald 2003; Nieto and Santamaría 2007).

10.3.2.3 Control Variables

As control variables, we considered those typical of studies on innovation and cooperation such as: firm size, technological intensity of the sector, ownership structure and innovation effort. The description of the measures used for these variables is shown in the Table 10.1.

10.3.3 Methodology

To check the hypotheses proposed, as the dependent variables are dichotomous, two independent probit models could be used, one for organizational innovations and one for marketing innovations. Nevertheless, it is to be expected that the error terms of these models considered together are correlated, which makes it more convenient to use an extended probit model, known as a bivariate probit (Greene 2000), which also allows us to consider the existence on unobservable factors influencing these decisions.

This econometric model has previously been applied by other authors in the field of innovation, more specifically in the study of the relationship between collaboration and different aspects of innovation activity, such as regularity in the

Table 10.1 Description of the control variables and their measurements

Variable	Measurement	Description
Technological intensity of the sector	High intensity (manufacturing)	1 if the firm belongs to a manufacturing sector of high technological intensity 0 if not
	Medium-high intensity (manufacturing)	1 if the firm belongs to a manufacturing sector of medium-high technological intensity 0 if not
	Medium-low intensity (manufacturing)	1 if the firm belongs to a manufacturing sector of medium-low technological intensity 0 if not
	Low intensity (manufacturing)	1 if the firm belongs to a manufacturing sector of low technological intensity 0 if not
	High intensity (services)	1 if the firm belongs to a service sector of high technological intensity 0 if not
	Low intensity (services)	1 if the firm belongs to a service sector of low technological intensity 0 if not
Size	Sales figures	Natural logarithm of the sales figure for 2009
Ownership structure	Private Spanish company	1 if the firm is a private Spanish company 0 if not
Innovation effort	Innovation expenses	Total figure of expenditure on innovation for the year 2006

performance of in-house R&D activities (Becker and Dietz 2004), the degree of novelty of the innovation developed (Nieto and Santamaría 2007), participation in national or international innovation programmes (Busom and Fernández-Ribas 2008), the effects of cooperation on the different types of innovation developed (Sánchez-González et al. 2008) or the impact of the different sources of knowledge on the results of the innovation process (Frenz and Ietto-Gillies 2009).

Two equations are considered in the specification of this model (Breen 1996):

$$
\begin{aligned}
y^*_1 &= \beta_1 x_1 + \varepsilon_1; y_1 = 1 si\, y^*_1 > 0, y_1 = 0 si\, y^*_1 \le 0 \\
y^*_2 &= \beta_2 x_2 + \varepsilon_2; y_2 = 1 si\, y^*_2 > 0, y_2 = 0 si\, y^*_2 \le 0 \\
&(\varepsilon_1, \varepsilon_2) \sim BVN(0,0,1,1,\rho)
\end{aligned}
$$

where y^*_1 and y^*_2 are latent variables, while y_1 and y_2 represent the dummy variables referring to the obtaining, respectively, of organizational and marketing innovations, β_1 and β_2 are the coefficients estimated for each of the two equations, x_1 and

x_2 the set of independent variables for each model and ε_1 and ε_2 the error terms following a distribution function of a bivariate normal whose correlation is determined by ρ.

Therefore, the model is constructed on the basis of two independent probit equations that can be estimated separately. However, as pointed out above, to ascertain whether it is appropriate to apply a bivariate probit, it is necessary to analyse the correlation between the error terms of the two equations and see if it is statistically significant. If it is not, it would be more appropriate to estimate each of the equations separately using separate (univariate) probits, for in such cases the bivariate probit would be less efficient (Greene 2000). For this purpose we use the Lagrange test, which operates under the null hypothesis that ρ equals zero.

On the other hand, the interpretation of these coefficients is not as simple as in a linear regression model, so it is necessary to analyse the marginal effects by calculating the change in the probability of obtaining one or the other type of innovation derived from a unitary increase in the explicative variables.

10.4 Results

Table 10.2 presents the results of the bivariate probit, where models 1 and 2 are, respectively, organizational and marketing innovations. As stated in the previous section, the variables referring to cooperation with different agents and the innovation effort have been delayed a period to measure their effect on the development of the two types of innovations being considered.

In the first place, to determine whether the bivariate probit model should be applied, we analysed the correlation between the error terms of the two equations to check its statistical significance. The LR test on the parameter ρ indicates that the correlation between the error terms of the two equations is statistically significant, the bivariate probit being the correct specification. Furthermore, according to the result of Wald's test, it may be stated that the set of variables selected are significant for both models.

On the basis of the results of the bivariate probit model, we can state that all the hypotheses are fulfilled except one, as there is a significant positive relationship between the collaboration with each of the different agents and the development of organizational and marketing innovations in all cases. Hypothesis H5 is the only one that is not fulfilled, as it proposed that cooperation with competitors had a negative influence on the development of organizational innovations, but the opposite result was obtained.

Specifically, we can say that hypotheses H1 and H2, relating to cooperation with suppliers, were both fulfilled, as these agents had a positive and significant influence on the probability of both organizational and marketing innovations being developed ($\beta = 0.299$ and $\beta = 0.306$ respectively). This result may be based on the ease of access to information on markets and experiences that suppliers may provide for the development of strategies and innovation plans. Furthermore, this relationship is

Table 10.2 Bivariate probit used to analyse the effect of cooperation with different agents on organizational and marketing innovations

		Model 1		Model 2	
Variables		Coeff.	dy/dx	Coeff.	dy/dx
Constant		−2.005***	–	−1.874***	–
		(0.115)	–	(0.121)	–
Explicative Variables					
Cooperation with suppliers		0.299***	0.118***	0.306***	0.105***
		(0.044)	(0.017)	(0.017)	(0.016)
Cooperation with customers		0.220***	0.087***	0.093*	0.031*
		(0.052)	(0.021)	(0.021)	(0.018)
Cooperation with competitors		0.145**	0.057**	0.098*	0.032*
		(0.060)	(0.024)	(0.024)	(0.020)
Cooperation with experts		0.231***	0.091***	0.101*	0.033*
		(0.053)	(0.021)	(0.021)	(0.018)
Cooperation with universities		0.233***	0.092***	0.207***	0.069***
		(0.043)	(0.017)	(0.017)	(0.015)
Control Variables					
Size		0.111***	0.043***	0.063***	0.020***
		(0.007)	(0.002)	(0.003)	(0.002)
Innovation effort		4.92e-09**	1.92e-09**	3.12e-09**	1.00e-09**
		(2.27e-09)	(0.000)	(0.000)	(0.000)
Technological intensity of the sector	High manufacture	0.133**	0.052**	0.247***	0.085***
		(0.056)	(0.022)	(0.022)	(0.020)
	Medium-low manufacture	−0.110***	−0.042***	−0.095**	−0.030**
		(0.037)	(0.014)	(0.014)	(0.012)
	Low manufacture	−0.161***	−0.062***	0.178***	0.059***
		(0.038)	(0.014)	(0.014)	(0.013)
	Low services	−0.390***	−0.145***	−0.277***	−0.083***
		(0.038)	(0.013)	(0.013)	(0.012)
Capital structure	Private Spanish	0.010	0.004	0.177***	0.054***
		(0.037)	(0.015)	(0.015)	(0.012)

Wald test $\chi2$ (24) = 1,028.18
Log likelihood = −11,845.798
N = 10,735
LR test on rho = 0
Value $\chi2$ (1) = 1,860.26 (0.000)

*$p<0.1$; **$p<0.05$; ***$p<0.01$
Note: Models 1 and 2 refer, respectively, to organizational and marketing innovations
Reference variables: sectors with a high-medium technological intensity in manufacturing and a high intensity in service firms
dx/dy estimates the discrete change of a dichotomous variable from 0 to 1
Standard errors in brackets

favoured by the special interest that suppliers may have in strengthening ties with their customers and increasing commercial activity of both through the positive effects that relationships of this type have on organizational as well as commercial activities.

Hypotheses H3 and H4 on the influence of cooperation with customers are also confirmed, as there is a significant and positive relationship between the variables ($\beta = 0.220$ and $\beta = 0.093$ respectively). These results may be explained bearing in mind that establishing relations with customers entails adapting the organization through the implementation of new ways of facilitating communication with them (organizational innovations). As for marketing innovations, they are favoured by cooperation with agents of this type as they facilitate rapid access to information useful for developing advertising campaigns, improvements in product presentation, promotions or new sales methods.

Regarding the relationship between cooperation with competitors and organizational innovations, we find the only hypothesis that is not confirmed. Hypothesis H5 proposed that this type of cooperation had a negative influence on the development of organizational innovations but the study shows there to be a significant and positive relationship between these variables ($\beta = 0.145$). This result may be explained by the fact that firms cooperating with competitors may have obtained information useful for imitating their competitors' organizational structure or management system, adapting them to their own needs. Hypothesis H6, on the other had, was confirmed, as there is a significant and positive relationship between cooperation with competitors and the development of marketing innovations, as initially proposed ($\beta = 0.098$). This means that firms working together with competitors have been able to jointly develop innovative communication strategies, such as the joint promotion of products or services in a city, region or country, or working together on launching new products or services developed in collaboration with several firms.

Hypotheses H7 and H8, on the effects of cooperation with experts and consultants, are also confirmed, as a significant and positive relationship is obtained for the two types of innovation ($\beta = 0.231$ and $\beta = 0.101$). From these results it may be understood that the help, experience and a different viewpoint that these agents may provide regarding non-technological issues foment the development of organizational innovations (such as the design of new structures or ways of managing human resources), and also innovations in marketing (like improvements to the product image or innovations in advertising and other novel commercial strategies).

The last hypotheses, H9 and H10, which analyse the effect of cooperation with universities and public research centres on non-technological innovations are also confirmed ($\beta = 0.233$ and $\beta = 0.207$ for organizational and marketing innovations, respectively). These results enable us to maintain that cooperation with these partners does not only contribute to developing technological innovations but also permits access to technical information for improving how work and staff are organized, developing new communication strategies or sales channels, and the recruitment of new talents or creative staff.

On the other hand it is important to point out that from the analysis of the marginal effects, it appears that collaboration with suppliers has the greatest effect on the development of innovations of both types, as such cooperation increased the probability of obtaining organizational and marketing innovations by, respectively 11.8 % and 10.5 % points, while everything else remained constant. Cooperation with the remaining agents, though with its significant and positive influence, achieves this in a proportion lower than 10 % points on the probability of developing innovations of both types, *ceteris paribus*. This indicates that firms must take the maximum advantage of relations with suppliers in the development of these innovations, given that with them, the relationship is usually closer and ongoing, which makes it easier to establish and maintain cooperation strategies (De Faria et al. 2010).

Comparing the marginal effects of the two models shows that for the five different types of agent, the influence of cooperation on innovation is greater in organization than in marketing. This may be because to achieve innovative changes in the organizational structure and management of a firm requires a deeper understanding of these issues and therefore a greater integration with cooperating partners. This is not the case, however, for marketing innovations, although cooperation does increase the probability of achieving them, but the relationship does not have to be so close.

As for the control variables, it is to be observed that there is a significant and positive relationship between size and the development of both organizational and marketing innovations. This is because the greater a firm's size, the more resources it has for the development of innovations, not only technological ones, such as in the product and process, but also non-technological ones. Likewise, there is a significant and positive relationship regarding innovation effort, which is to be expected, for the greater the investment in innovation, the greater the probability of any kind of innovation actually being achieved.

Regarding the technological intensity of the sector, differences exist between manufacturing and service firms. The former show a different behaviour for innovations of these two kinds according to their technological intensity. As for organizational innovations, belonging to sectors with a low technological intensity has a significant and negative effect on their development while belonging to sectors with a high technological intensity exerts a positive and significant effect (low intensity $\beta=-0.161$, medium-low intensity $\beta=-0.110$ and high intensity $\beta=0.133$). These results are due to firms' need to adapt their organizational structure to the technological changes that they make (Dougherty 1992; Danneels 2002). Therefore, the greater the technological intensity of their sector is, the greater their need to innovate in general and the greater the need for organizational changes in particular will be. In contrast, in the case of marketing innovations, although there is a significant relationship, the direction is not so clear (low intensity $\beta=0.178$, medium-low intensity $\beta=-0.095$ and high intensity $\beta=0.247$), which could be due to the current competitive environment obliging firms to innovate continually in marketing to favour sales, regardless of the sector a firm belongs to. In the case of service firms, we observe that belonging to sectors with a low technological intensity has a significant and negative influence on both types of non-technological innovation ($\beta=-0.390$ and $\beta=-0.277$). Unlike manufacturing firms, service firms with a high technological level seem to be more motivated to develop both types of innovation. These results may be due to service companies being characterized by the need for much closer contact with the market than manufacturers. Services with a low technological content are widely known in the market and do not need to make so much commercial effort. On the other hand, service firms with a high technological level are up against more competitive and dynamic environments. This obliges them to be continually innovating in the services that they offer, with the consequent need for the ongoing adaptation of their organizational structure and in their way of presenting themselves and informing of their services in the market.

Finally, an analysis of capital structure shows that there is no significant relationship between being a private Spanish company and the development of organizational innovations. We may therefore conclude that capital structure does not play an important role in the development of innovations of this type. However, being a Spanish company does have a significant and positive influence on the development of innovations in marketing. This is because, if the firm is Spanish, it has the need to internally develop its own strategies for selling its products while belonging to a multinational means that such strategies emanate from the parent company.

10.5 Conclusions

This paper concentrates on the study of the effects of cooperation with different types of external agents on the development of non-technological innovations. To this end, we have used data from the Panel of Technological Innovation and designed a bivariate probit model with two types of non-technological innovations as dependent variables: (1) organizational innovations and (2) marketing innovations. With this model we analysed the influence of cooperation with five types of external agents: (1) suppliers, (2) customers, (3) competitors, (4) experts and consultants and (5) universities and public research centres, and the relative importance of each of them.

The results lead to three important conclusions. Firstly, it may be stated that cooperation with external agents favours the development of both organizational and marketing innovations. Secondly, suppliers are demonstrated to be the external agent that contributes most to the development of these innovations. Thirdly, cooperation with different types of agent has a greater positive effect on the development of organizational innovations than marketing ones.

In a complementary way, the results also show that size favours the development of both types of non-technological innovations, and that the greater the firm's innovation effort, the greater the probability of achieving any kind of innovation. Regarding the technological intensity of sectors, manufacturing firms belonging to sectors with a high technological level have the greatest probability of developing organizational innovations. For marketing innovations, however, this relationship is not clear. In contrast, service firms with a high technological intensity develop both types of innovation equally. Finally, regarding capital structure, the fact of being a private Spanish company has no influence on the development of organizational innovations but it does in the case of marketing ones.

This study contributes to the previously existing literature on the influence of cooperation on the development of non-technological innovations by tackling the subject in such a way that, to our knowledge, is novel. Thus, it may be said that the study contributes empirical evidence that goes beyond the scope of other studies, which merely analyse the effects of these relationships on the development of innovations in general or only technological ones (Sánchez-González et al. 2008; Frenz and Ietto-Gillies 2009; Tödtling et al. 2009; De Faria et al. 2010; Tomlinson 2010; Laursen 2011). It also complements the results of previous studies on non-technological innovations that do not consider the effects of cooperation (Dengbo et al. 2008; Nguyen and Mothe 2008; Evangelista and Vezzani 2010; Naidoo 2010; Afcha Chávez 2011) and others that, while bearing in mind the effects on non-technological innovations, do not analyse them by type of agent (Armbruster et al. 2008; Yang et al. 2008; Camisón and Villar-López 2011; Baraldi et al. 2012).

Therefore, the fundamental idea deriving from this study is that cooperation, regardless of the type of partner, is a good strategy for firms to use with the aim in view of developing non-technological innovations. It would therefore be recommendable for firms to work in an integrated way collaborating with a wide number of agents, especially when they wish to perform structural or management changes in the organization and, though with less importance, when they wish to design innovative marketing strategies. To propose ideas to carry out innovations of an organizational type requires the external agent to know the structure of the firm in depth, along with its ways of organizing itself and working, so in these cases, a greater integration of the partner in the firm's activities is necessary. Likewise, although the five types of agent can make valuable contributions in this context, suppliers are the most influential ones. Their importance is due to the closer and more frequent ties that firms have with their suppliers, which generates conditions of trust and exchange of knowledge that are more difficult to establish with other agents and which may facilitate the development of joint activities.

Finally, it is important to point out certain limitations of the study for a correct interpretation of the results and conclusions, which also give rise to future lines of research. In the first place, the database used did not make it possible to perform an analysis by years, given that the questions in the questionnaire, both on cooperation and on obtaining innovations are made for a period of 3 years. Another limitation is that both the dependent and explicative variables were measured by means of dichotomous variables, and without a doubt, a greater contribution would have been possible using measurements on the intensity of cooperation and of the success of the ensuing innovations. Further work on these questions would allow us to draw father-reaching conclusions and analyse the phenomenon in greater depth.

Furthermore, the database used only contains information on Spanish firms, so the conclusions only concern Spain. Future research would enrich our knowledge by using data from other countries in order to make comparisons. On the other hand, it would be interesting to analyse the effects of cooperation by distinguishing organizational and marketing innovations according to the type of activities comprising them, and study the technological innovations (in product and process) and non-technological ones together in order to complete the study analysing their interaction.

References

Afcha Chávez SM (2011) Innovaciones organizacionales y su efecto sobre el desempeño empresarial. Revista Venezolana de Gerencia 56(16):544–563

Ahuja G (2000) The duality of collaboration: inducements and opportunities in the formation of interfirm linkages. Strateg Manag J 21(3):317–343

Armbruster H, Kirner E, Lay G (2006) Patterns of organisational change in European industry (PORCH). Ways to strengthen the empirical basis of research and policy. Fraunhofer Institute of Systems and Innovation Research, Karlsruhe

Armbruster H, Bikfalvi A, Kinkel S, Lay G (2008) Organizational innovation: the challenge of measuring non-technical innovation in large-scale surveys. Technovation 28:644–657

Azagra-Caro JM, Archontakis F, Gutierrez-Gracía A, Fernández-de-Lucio I (2006) Faculty support for the objectives of university–industry relations versus degree of R&D cooperation: the importance of regional absorptive capacity. Res Policy 35:37–55

Baker WE, Sinkula JM (1999) The synergistic effect of market orientation and learning orientation. J Acad Mark Sci 27(4):411–427

Baraldi E, Gressetvold E, Harrison D (2012) Resource interaction in inter-organizational networks: introduction to the special issue. J Bus Res 65:123–127

Bayona C, García-Marco T, Huerta E (2000) Situación de la cooperación en I+D en España con Universidades y centros de investigación. Documento de trabajo, Universidad Pública de Navarra

Bayona C, García-Marco T, Huerta E (2001) Firm's motivations for cooperative R&D: an empirical analysis of Spanish firms. Res Policy 30:1289–1307

Bayona C, García-Marco T, Huerta E (2003) ¿Cooperar en I+D? Con quién y para qué. Revista de Economía Aplicada 31(11):103–134

Becker W, Dietz J (2004) R&D cooperation and innovation activities of firms: evidence for the German manufacturing industry. Res Policy 33:209–223

Belderbos R, Carree M, Lokshin B (2004a) Cooperative R&D and firm performance. Res Policy 33:1477–1492

Belderbos R, Carree M, Diederen B, Lokshin B, Veugelers R (2004b) Heterogeneity in R&D cooperation strategies. Int J Ind Organ 22:1237–1263

Bessant J, Rush H (1995) Building bridges for innovation: the role of consultants in technology transfer. Res Policy 24:97–114

Biemans WG (1991) User and third-party involvement in developing medical equipment innovations. Technovation 11:163–182

Black SE, Lynch LM (2004) What's driving the new economy? The benefits of workplace innovation. Econ J 114(493):97–116

Bogers M, Afuah A, Bastian B (2010) Users as innovators: a review, critique and future research directions. J Manag 36(4):857–875

Breen R (1996) Censored, sample-selected, or truncated data regression models. Sage, Beverly Hills

Bruce M, Morris B (1998) In house, out-sourced or a mixed approach to design. In: Bruce M, Jevnaker B (eds) Management of design alliances: sustaining competitive advantage. Wiley, Chichester

Burnham JB (1997) Evaluating industry/university research linkages. Res-Technol Manag 40:52–55

Busom I, Fernández-Ribas A (2008) The impact of firm participation in R&D programmes on R&D partnerships. Res Policy 37:240–257

Camisón C, Villar-López A (2011) Non-technical innovation: organizational memory and learning capabilities as antecedent factors with effects on sustained competitive advantage. Ind Mark Manag 40:1294–1304

Cassiman B, Veugelers R (1998) R&D cooperation and spillovers: some empirical evidence. Documento de trabajo nº 328, Universitat Pompeu Fabra

Cassiman B, Veugelers R (2002) R&D cooperation and spillovers: some empirical evidence from Belgium. Am Econ Rev 92(4):1169–1184
Cassiman B, Veugelers R (2005) R&D cooperation between firms and universities. Some empirical evidence from Belgian manufacturing. Int J Ind Organ 23(5–6):355–379
Clark KB (1989) Project scope and project performance: the effect of parts strategy and supplier involvement on product development. Manag Sci 35:1247–1263
Cohen WM, Nelson RR, Walsh J (2002) Links and impacts: the influence of public research on industrial R&D. Manag Sci 48:1–23
Conway S (1995) Informal boundary-spanning communication in the innovation process: an empirical study. Technol Anal Strateg Manag 7:327–342
Damanpour F, Aravind D (2012) Organizational structure and innovation revisited: from organic to ambidextrous structure. In: Mumford M (ed) Handbook of organizational creativity. Elsevier, Oxford, pp 483–515
Danneels E (2002) The dynamics of product innovation and firm competencies. Strateg Manag J 23:1095–1121
Day G (1994) The capabilities of market-driven organizations. J Mark 58(4):37–43
De Faria P, Lima F, Santos R (2010) Cooperation in innovation activities: the importance of partners. Res Policy 39:1082–1092
Deák C (2006) From innovation management to innovation marketing – knowledge management by the innovation management cooperation research center. ISPIM international conference, Athens
Dengbo Z, Wulamu A, Lixin G (2008) Research of methodology and theory of non-technical innovation. 4th international conference on wireless communications, networking and mobile computing, WiCOM '08
Deshpandé R, Farley JU, Webster FE (1993) Corporate culture, customer orientation and innovativeness in Japanese firms: a quadrat analysis. J Mark 57:23–37
Dougherty D (1992) Interpretive barriers to successful product innovations in large firms. Organ Sci 3(2):179–202
Duysters G, Kok G, Vaandrager M (1999) Crafting successful strategic technology partnerships. R&D Manag 29(4):343–351
Edquist D (1997) Systems of innovation. Technologies, institutions and organizations. Pinter/Cassell, London
Edquist D, Hommen I (1999) Systems of innovation: theory and policy for the demand side. Technol Soc 21:63–79
Erickson C, Jacoby S (2003) The effects of employer networks on workplace innovation and training. Ind Labor Relations Rev 56:203–223
Evangelista R, Vezzani A (2010) The economic impact of technological and organizational innovations. A firm-level analysis. Res Policy 39:1253–1263
Fontana R, Geuna A, Matt M (2003) Firm size and openness: the driving forces of university–industry collaboration. SPRU electronic working paper, SEWP 103. University of Sussex, Brighton
Fontana R, Geuna A, Matt M (2006) Factors affecting university–industry R&D projects: the importance of searching, screening and signalling. Res Policy 35:309–323
Foss NJ, Laursen K, Pedersen T (2011) Linking customer interaction and innovation: the mediating role of new organizational practices. Organ Sci 22(4):980–999
Frenz M, Ietto-Gillies G (2009) The impact on innovation performance of different sources of knowledge: evidence from the UK community innovation survey. Res Policy 38:1125–1135
Fritsch M, Lukas R (2001) Who cooperates on R&D? Res Policy 30:297–312
Fujimoto T, Iansiti M, Clark KB (1996) External integration in product development. In: Nishiguchi T (ed) Managing product development. Oxford University Press, Oxford, pp 121–161
Garrette B, Doussauge P (1995) Les stratégies d´alliance. Les Editions d´Organisation, Paris
Gemünden HG, Heydebreck P, Herden R (1992) Technological interweavement: a means of achieving innovation success. R&D Manag 22(4):359–375
Gera S, Gu W (2004) The effect of organizational innovation and information and communications technology on firm performance. Int Prod Monit 9:37–51

Greene W (2000) Econometric analysis. Prentice-Hall, New Jersey

Hagedoorn J (1996) Trends and patterns in strategic technology partnering since the early seventies. Strateg Manag J 14:371–385

Hagedoorn J (2002) Inter-firm R&D partnership: an overview of major trends and patterns since 1960. Res Policy 31:477–492

Hakanson L (1993) Managing cooperative research and development: partner selection and contract design. R&D Manag 23(4):273–285

Holbrook JA, Wolfe DA (2000) Innovation, institution and territory. Regional innovation systems in Canada. McGill-Queen's University Press, Kingston

Hurley RF, Hult GTM (1998) Innovation, market orientation, and organizational learning: an integration and empirical examination. J Mark 62(3):42–54

Jeppesen LB (2002) Making consumer knowledge available and useful. The case of the computer games. DRUID working paper no 01–10, 2ª version

Jordá Borrell R (2005) Modificación del entorno y proceso innovador de las empresas de servicios avanzados en Andalucía. Investigaciones Geográficas 37:73–92

Kleinknecht A, Reijnen J (1992) Why do firms cooperate on R&D? An empirical study. Res Policy 21:347–360

Lam A (2005) Organizational innovation. In: Fagerberg J, Mowery DC, Nelson RR (eds) The Oxford handbook of innovation. Oxford University Press, Oxford, pp 115–147

Landry R, Amara N (2001) Creativity, innovation and business practices in the matter of knowledge management. In: De la Mothe J, Foray D (eds) Knowledge management in the innovation process: business practices and technology adoption. Kluwer Academic Publishers, Boston, pp 55–80

Laursen K (2011) User–producer interaction as a driver of innovation: costs and advantages in an open innovation model. Sci Public Policy 38(9):713–723

Leonard D, Rayport J (1997) Spark innovation through empathic design. Harv Bus Rev 75(6): 102–113

Leyden DP, Link AN (1992) Government's role in innovation. Kluwer Academic Publishers, Boston

Leyden DP, Link AN (1999) Federal laboratories as research partners. Int J Ind Organ 17:557–574

Lilien G, Morrison PD, Searls K, Sonnack M, von Hippel E (2002) Performance assessment of the lead user generation process for new product development. Manag Sci 48(8):1042–1060

Link AN, Scott JT (2005) Universities as partners in U.S. research joint ventures. Res Policy 34:385–393

López A (2008) Determinants of R&D cooperation: evidence from Spanish manufacturing firms. Int J Ind Organ 26(1):113–136

Lorenzoni G, Lipparini A (1999) The leveraging of interfirm relationships as a distinctive organizational capability: a longitudinal study. Strateg Manag J 20:317–338

Lundvall B-Å (1992) User–producer relationships, national systems of innovation and internationalisation. In: Lundvall B-Å (ed) National systems of innovation. Pinter, London, pp 45–67

Mariti P, Smiley RH (1983) Co-operative agreements and the organization of industry. J Ind Econ 31(4):437–451

Meyer-Krahmer F, Schmoch U (1998) Science-based technologies: university–industry interactions in four fields. Res Policy 27:835–851

Miotti L, Sachwald F (2003) Co-operative R&D: why and with whom? An integrated framework of analysis. Res Policy 32:1481–1499

Mowery DC, Oxley JE, Silverman BS (1998) Technological overlap and interfirm cooperation: implications for the resource base view of the firm. Res Policy 27:507–523

Murphy M (2002) Organisational change and firm performance. OECD science, technology and industry working papers

Naidoo V (2010) Firm survival through a crisis: the influence of market orientation, marketing innovation and business strategy. Ind Mark Manag 39:1311–1320

Neale MR, Corkindale DR (1998) Co-developing products: involving customers earlier and more deeply. Long Range Plan 31(3):418–425

Nguyen T, Mothe C (2008) Assessing the impact of marketing and organizational innovations on firm performance. International business information management conference, Kuala Lumpur
Nieto MJ, Santamaría L (2007) The importance of diverse collaborative networks for the novelty of product innovation. Technovation 27(6–7):367–377
Nishiguchi T, Ikeda M (1996) Suppliers' innovation: understated aspects of Japanese industrial sourcing. In: Nishiguchi T (ed) Managing product development. Oxford University Press, Oxford, pp 206–232
OECD (2005) Oslo manual. Guidelines for collecting and interpreting innovation dates, 3rd edn. Paris
Ritter T, Gemünden HG (2003) Interorganizational relationships and networks: an overview. J Bus Res 56:691–697
Robertson M, Swan J (1996) The role of networks in the diffusion of technological innovation. J Manag Stud 33(3):333–359
Romijn H, Albaladejo M (2002) Determinants of innovation capability in small electronics and software firms in southeast England. Res Policy 31(7):1053–1067
Romijn H, Albu M (2002) Innovation, networking and proximity: lessons from small high technology firms in the UK. Reg Stud 36:81–86
Rust RT, Lemon KN, Zeithaml VA (2004) Return on marketing: using customer equity to focus marketing strategy. J Mark 68(1):109–127
Sánchez-González G, González-Álvarez N, Nieto M (2008) Efectos de la colaboración con usuarios sobre el tipo de innovación desarrollado. Innovar: J Adm Social Sci 18(32):87–110
Shaw B (1994) User–supplier links and innovation. In: Dodgson M, Rothwell R (eds) The handbook of industrial innovation. Edward Elgar, Cheltenham
Tether B (2002) Who cooperates for innovation and why. An empirical analysis. Res Policy 31:947–967
Tödtling F, Lehner P, Kaufmann A (2009) Do different types of innovation rely on specific kinds of knowledge interactions? Technovation 29:59–71
Tomlinson PR (2010) Co-operative ties and innovation: some new evidence for UK manufacturing. Res Policy 39:762–775
Uhlaner L, Van Stel A, Meijaard J, Folkeringa M (2007) The relationship between knowledge management, innovation and firm performance: evidence from Dutch SMEs. Scientific analysis of entrepreneurship and SMEs (report). pp 1–26
Urban GL, von Hippel E (1988) Lead user analysis for development of new industrial products. Manag Sci 34(5):569–582
Veugelers R, Cassiman B (2005) R&D cooperation between firms and universities. Some empirical evidence from Belgian manufacturing. Int J Ind Organ 23:355–379
von Hippel E (1987) Cooperation between rivals: informal know-how trading. Res Policy 16:291 302
von Hippel E (1988) The sources of innovation. Oxford University Press, New York
von Hippel E, Katz R (2002) Shifting innovation to users via toolkits. Manag Sci 48(7):821–833
Wagner SM, Hoegl M (2006) Involving suppliers in product development: insights from R&D directors and project managers. Ind Mark Manag 35(8):936–943
Yang J, Wang S, Wu L, Lan H (2008) An empirical study on the relationships among inter-firm cooperation, knowledge creation and organizational innovation. The seventh Wuhan international conference on e-business: unlocking the full potential of global technology. Proceedings, 31 of May – 1 of June, 2008, Wuhan

Chapter 11
Management Innovation Strategy: Patterns, Antecedents and Synchronous Co-adoption

José-Luis Hervás-Oliver, Francisca Sempere-Ripoll, and Marta Peris-Ortiz

Abstract This paper sheds light on the antecedents of management innovation adoption in firms, exploring the simultaneous co-adoption or concurrence with technological innovations. Analyzing 12,824 firms using Spanish CIS 2006 data, we elaborate on previous models (Ganter and Hecker (J Bus Res 66(5):575–584, 2012); Mol and Birkinshaw (J Bus Res 62(12):1269–1280, 2009)), extending them by incorporating two new elements: (1) a resource-based view framework and, (2) the complementarities and organizational integration generated from the simultaneous co-adoption of management and technological innovations. Our results partially confirm those found in the previous models but extend them by showing that management innovation is frequently accompanied by technological innovation, especially when firms are complex innovators. Significant cross-country differences arise, based on institutional environment variety.

11.1 Introduction

The term *management innovation* encompasses the introduction of new administrative (e.g. Kimberly and Evanisko 1981), organizational (e.g. Armbruster et al. 2008) and managerial (e.g. Birkinshaw et al. 2008) activities, although currently it is accepted that all these terms overlap (e.g. Damanpour and Aravind 2011: 35). In this paper we will focus on the two types of management innovation defined by the Oslo Manual (2005), that is, organizational and marketing. Following the Oslo Manual (OECD 2005: 51), *organizational* innovation is recognized as "*the implementation of a new organizational method in the firm' business practices,*

J.-L. Hervás-Oliver (✉) • F. Sempere-Ripoll • M. Peris-Ortiz
Department of Business Organization, Universitat Politècnica de València, Valencia, Spain
e-mail: jose.hervas@doe.upv.es; fsempere@omp.upv.es; mperis@doe.upv.es

J.-L. Hervás-Oliver and M. Peris-Ortiz (eds.), *Management Innovation*,
Springer Proceedings in Business and Economics, DOI 10.1007/978-3-319-03134-7_11,

© Springer International Publishing Switzerland 2014

workplace organization or external relations (…) that has not been used before in the firm and is the result of strategic decisions taken by management". Similarly, marketing innovation defined as the "A marketing innovation is the implementation of a new marketing method involving significant changes in product design or packaging, product placement, product promotion or pricing"(pp. 49).

Under this framework, in which technology innovation is markedly separated from the management one, this work advances knowledge on the topic of management innovation by exploring the antecedents of a firm's adoption of management innovation. Specifically, this study elaborates on the recent work of Mol and Birkinshaw (2009), based at UK CIS 3 (1998–2000), and the subsequent replication by Ganter and Hecker (2012) in Germany, based at CIS 4 (2003–2005). This study is a critical testing of those works, in order to validate models and theories in cross-country comparisons. In addition, our work substantially extends those works by considering *complementarities* (Milgrom and Roberts 1995) or a *synchronous adoption approach* (Ettlie 1988) by the inclusion of technological innovation in the model. All in all, our work constitutes an attempt to provide a response to the call made by those researchers to find out more about organizational innovation in different countries. For this purpose, we use Spanish CIS 2006 data (2004–2006) in order to respond to those calls.

Studies about the adoption of management innovation adoption are increasing (e.g. Birkinshaw et al. 2008; Vaccaro et al. 2012; Walker et al. 2011). Our paper makes a contribution to the debate by understanding the influence of technology innovation on a firm's co-adoption of management innovation, that is, management innovation is adopted *in tandem* with the technological (product and process) modes. Our paper builds a more comprehensive model in which to place the issue of management innovation together with technological innovation. In addition, this paper utilizes a complementary framework to the *context* and *search* perspective seen in Mol and Birkinshaw, using the resource-based view (RBV) of the firm (e.g. Barney 1991; Wernerfelt 1984). This cross-theory analysis allows a more interesting debate and permits researchers to assess and discuss results using distinct but complementary frameworks.

Overall, this paper contributes to the literature in the following ways. Firstly, the paper provides insights about the antecedents of the adoption of management innovation, assessing the few existing works (such as Mol and Birkinshaw and Ganter and Hecker – which we call *baseline* articles). As a result, a better theoretical framework about management innovation can be built and knowledge about the topic extended, enriching the innovation literature. Secondly, the paper also contributes to the literature on management innovation by exploring the latter's complementary role with technological innovation, providing thus a more robust and better theoretical connection between technological and management innovation. Thirdly, the work discusses management innovation using the RBV, enriching thus that theoretical approach. The paper is organized as follows. In Sect. 11.2 we make a revision of the theory and extend the stated models following a RBV approach. Section 11.3 presents the empirical design. Then, in the fourth section results are presented and discussed. Finally, Sect. 11.5 concludes the analysis and summarizes the theoretical contributions and implications.

11.2 Theoretical Framework

11.2.1 Using the RBV Framework to Explain Antecedents for Management Innovation

Internal and external sources of knowledge are constituents of a firm's resources and capabilities (Barney 1991), or organizational capabilities (Grant 1996a), the formation of which, the way they combine, and the way they develop, make up an organizational learning platform (e.g. Dodgson 1993; Garvin 1993). The output of the latter is organizational knowledge, which is the antecedent of innovation (Cohen and Levinthal 1990; Coombs and Hull 1998; Crossan and Apaydin 2010; Grant 1996b; e.g. Kimberly and Evanisko 1981; Kogut and Zander 1992; Lam 2004, 2005).

In general, the organizational learning literature, alike with the innovation one, suggests different models to express the relationship between organizational learning and the innovation function (e.g. the absorptive capacity, by Cohen and Levinthal 1990; the knowledge management, Nonaka and Takeuchi 1995, among others). In this chain of thought, organizational capabilities mainly refer to the organizational rules, procedures and values that are involved in the coordination of functional capabilities, and in the cohesion of the members of the organization (e.g. Nelson and Winter 1982). That construct is often operational using the dynamic capabilities (e.g. Teece et al. 1997) view and the resource-based view (RBV, e.g. Barney 1991). Barney (1991) referred to the RBV (using a broad definition of resources) as all types of assets, organizational processes, knowledge capabilities and other potential sources of advantage.

All in all, investment in resources to help the development of organizational capabilities is necessary to provide unique or rare capabilities Damanpour et al. (2009) which can trigger innovations, and improve performance through the building up of competitive advantage (Helfat et al. 2007). These internal and external resource combinations (e.g. Laursen and Salter 2006; Vega-Jurado et al. 2008) will endow a firm with better organizational knowledge or organizational capabilities to innovate (e.g. Cohen and Levinthal 1990). Organizational capabilities are formed by the combination of internal and external sources of knowledge in firms. A focus on these phenomena looks at organizational innovation not as an outcome, but, rather, in terms of the specific organizational capabilities required for innovation – that is, the precondition for ensuring innovation in organizations thanks to appropriate and key organizational characteristics which enhance a firm's capacity for innovation. Internal resources include the education quality of the workforce (Alvesson 1995; Chandler 1962; Daft 1978; Mol and Birkinshaw 2009). The internal resources relevant for promoting management innovation are associated with the administrative and managerial practices of people in organizations. This idea is also described in different perspectives. For instance, the socio-technical system (Trist and Bamforth 1951) describes a method of describing organizations, emphasizing the interrelatedness of the way social and technological subsystems function. The socio-technical perspective contends that organizations are made up of people and technology (Pasmore et al. 1982; Trist 1978). In this vein, management innovation adoption requires capabilities directly related to the social system or people in the organization, beyond the technological ones.

Complementing the RBV, the relational perspective(Dyer and Singh 1998)refers to the way external resources also nurture knowledge in firms and build up organizational capabilities. Those resources arise from collaboration (Cebon et al. 1999) with universities (Atuahene-Gima 1995), customers (Lee 1996), or suppliers (Bessant 2003), and this process of sourcing knowledge generates further relational returns (Dyer and Singh 1998) by facilitating innovation through knowledge-sharing and an interactive learning process (Lee et al. 2001; McEvily and Zaheer 1999; Powell et al. 1996; Rowley et al. 2000) which enriches a firm's repository of knowledge or organizational capabilities. In this vein, the literature has pointed out that the greater the breadth of knowledge sourced by firms, the greater the introduction of new management practices (Birkinshaw et al. 2008; Ganter and Hecker 2012; Mol and Birkinshaw 2009). Following Damanpour et al. (2009) active search in both internal and external sources of knowledge positively affects the innovation process. In particular, for the specific case of management innovation, a variety of external actors provide knowledge and new ideas which then lead to the introduction of new practices (Abrahamson 1996; Guler et al. 2002; Staw and Epstein 2000), while internal communication among the members of an organization facilitates the dispersion of ideas which then increase in amount and diversity, resulting in a cross-fertilization of ideas (Aiken and Hage 1971), the creation of an internal environment favourable to the survival of new ideas (Ross 1974), and the introduction of new management practices (e.g. Hansen and Løvås 2004; Nahapiet and Ghoshal 1998). All in all, it is expected that, ceteris paribus, the greater the amount of investment in external and internal sources of knowledge, the more are built innovation-based organizational capabilities which foster the introduction of new management practices. Thus, the RBV and relational view framework presented is equivalent to the Mol and Birkinshaw's organizational reference group (context) literature and search strategy approach. Once this paper has presented the RBV framework to address the key requirements for understanding the antecedents of management innovation, the basic models of Mol and Birkinshaw (2009) and Ganter and Hecker (2012) are presented and then elaborated upon. Larger firm size facilitates the adoption of organizational innovation(Damanpour 1987; Kimberly and Evanisko 1981). The reason is that increasing size demands innovative organizational methods to cope with arising coordination problems. We replicate the hypothesis.

Hypothesis 1: The larger the firm, the higher the level of adoption of new organizational innovation

The education of the workforce, measured as the percentage of employees with a degree, is also potentially an important attribute of the firm and represents one of its key innovation resources, to the extent that many organizational innovations require a high level of skills and education (e.g. Chandler 1962; Ichniowski et al. 1997). We replicate the hypothesis two from the baseline papers.

Hypothesis 2. The more highly educated the workforce of the firm, the greater the level of introduction of new management practices.

Participation in international markets may be a source of insight for management innovation, since it exposes firms to a much broader set of management approaches and opportunities in different contexts than they would experience in their domestic markets (e.g. Kogut and Parkinson 1993). Following the same hypothesis as the baseline papers:

Hypothesis 3. The greater the geographical scope of the market the firm is operating in, the greater the level of introduction of new management practices.

As outlined in the RBV framework, the Dyer and Singh (1998) relational perspective on learning and knowledge-sharing emphasises an interactive learning process (Lee et al. 2001; McEvily and Zaheer 1999; Powell et al. 1996; Rowley et al. 2000) which enriches a firm's repository of knowledge and organizational capabilities, and establishes an alternative repository of knowledge for introducing innovations (e.g. Laursen and Salter 2006; Singh 2005). We have distinguished between internal and external sources of knowledge, simplifying the original three hypotheses stated in Mol and Birkinshaw and Ganter and Hecker. The reason is that we do not want to provide an ex-ante classification or taxonomy of the external sources for promoting innovation, as they are not classified in the way that Mol and Birkinshaw follow in the CIS database they use. Nevertheless, our approach is equivalent to the baseline works.

Hypothesis 4a. The more internal sources the firm interacts with, the greater the level of introduction of new management practices.

Hypothesis 4b. The more external sources of knowledge the firm interacts with, the greater the level of introduction of new management practices.

Finally, the stated baseline works offer some interactive terms in respect of the internal sources of knowledge, or context construct, in Mol and Birkinshaw, and the external sources of knowledge from the search strategy. In this case, instead of looking for complementarities, that is, the positive effect from combining internal and external assets (e.g. Cassiman and Veugelers 2006; Escribano et al. 2009), Mol and Birkinshaw rely on a substitution effect (e.g. Laursen and Salter 2006) in which the organizational context (or internal capabilities) act as substitutes for knowledge-search (external sources or capabilities) or vice-versa. Specifically, in Mol and Birkinshaw the expected (and evidenced) negative relationship means that increasing size, workforce education and geographic scope makes firms less dependent on tapping into external sources of knowledge to innovate. In this vein, Mol and Birkinshaw found negative and significant coefficients which support the substitution effect. We, again, synthesise their hypothesis into the following (equivalent) ones:

Hypothesis 5a. The effect of internal sources on the introduction of new management practices is mitigated by size, education of the workforce, and geographic scope of the firm.

Hypothesis 5b. The effect of external sources of knowledge on the introduction of new management practices is mitigated by size, education of the workforce, and geographic scope of the firm.

11.2.2 Model Extension

The extension provided by Ganter and Hecker (2012) addressing environmental influences can only be partially replicated, due to the fact that CIS surveys present many cross-country differences, although they tend to be harmonized. For instance, in Spanish CIS, measures such as *product homogeneity*, *product life cycle* or *technological change* are not listed in the official survey .[1] Nevertheless, we think that the baseline model can be extended differently by introducing the idea of *complementarities* (Milgrom and Roberts 1995), together with the *synchronous adoption* of technological and management innovations (e.g. Ettlie 1988) or organizational integration (Ettlie and Reza 1992). Thus, product and process innovation, used as control variables in the baseline paper, come onto the scene as theoretical antecedents of the adoption of management innovation.

Damanpour et al. (2009: 651) point out that socio-technical system theory (Trist and Bamforth 1951) has challenged the technological imperative, arguing that changes in the technical system of the organization should be coupled with changes in the social (organizational) system in order to optimize organizational outcomes (Cummings and Srivastva 1977; Damanpour and Evan 1984). Therefore, the adoption of organizational innovation is an effective way of complementing and supporting technical innovation. Ettlie (1988) dubs the simultaneous use of management innovation and technological innovation "synchronous innovation", and argues that the use of appropriate forms of management innovation makes technological innovation more effective due to the existence of complementarities and synergies between them. The combined adoption of both modes is stated in Damanpour and Evan (1984), which claims that high-performing firms are the ones which, vis-à-vis low-performing ones, present a stronger association between technical and administrative (non-technical) innovations. The reason for this mutually beneficial co-adoption is the fact that *complementarities* arise. Milgrom and Roberts (1995) and Ichniowski et al. (1997)focus on the notion of complementarities as systemic changes in organizational practices. In the management literature, it is evidenced that there is a correlative relationship between organizational subsystems in the form of a "coupling of dissimilarities", where each change in a subsystem requires alterations in the other subsystems (Trist and Murray 1993). All in all, *organizational integration* (Ettlie and Reza 1992), following the introduction of new organizational activities aimed at integrating technological change, constitutes a ferment for technological innovations and induces a need for integrating other managerial and organizational innovations. The point is to understand that achievement of competitive advantage by building innovative capabilities requires building complex systems of activities in which the number of elements, and their interactions, makes for an inimitable system (Rivkin 2000). Thus, the extension of the model we propose, to those used as a baseline, is depicted in the following hypothesis:

[1] See more about the Spanish survey at: http://www.ine.es/jaxi/menu.do;jsessionid=C83F6769A6B193510E8686BD2B4C91D4.jaxi01?type=pcaxis&path=%2Ft14/p061&file=inebase&L=1.

Hypothesis 6: The introduction of new technological (product and process) innovations fosters the adoption of management innovation due to need to integrate technology into the organization

11.3 Empirical Design

11.3.1 Dataset, Sample, Variables and Methods

The method and the types of questions in CIS are described in the Oslo Manual (Oslo manual: OECD 2005). CIS data is used for the study of innovation at the firm level in a large number of studies, in countries such as Belgium, France, Spain and others (e.g. Crepon et al. 1998; Escribano et al. 2009; Evangelista and Vezzani 2010; Laursen and Salter 2006; Mol and Birkinshaw 2009; Tether 2002). Our final sample covers 12,824 firms, with complete information. The variables are showed in Table 11.1: dependent, independent and control variables. Most of the variables are those taken from the baseline articles, except some variables which are not available in the Spanish data. The empirical design follows, in part, that of Ganter and Hecker (2012). Each estimated model first uses an ordered logistic regression, following the strict definition of organizational innovation from the Oslo Manual, excluding marketing innovation. Then, we repeat the procedures, adding marketing innovation in the dependent constructs, following Mol and Birkinshaw's definition of the management innovation construct based on the Oslo Manual. In short, model A presents an ordered logit for organizational innovation (Inno_org_sum, 0–4) and B a binary logit for the same dependent variable (Inno_org, 0–1). Then, model C replicates A by adding marketing innovation (Inno_management_sum, 0–8), and D replicates B by adding marketing innovation (inno_management, 0–1). Eventually, the work uses as a control the industry classification, using 58 2-digit NACE-93 industry dummies, ranging from the 14 to 74 codes (59 industries or dummies). Code 55 is the baseline. Table 11.2 provides the main descriptive statistics. See Table 11.2.

As aforementioned, this paper avoids the ex-ante classification of external sources of knowledge in the search strategy (relational capabilities) approach. For this reason, the paper conducts an ex-post aggregation of those sources using Principal Component Analysis (PCA). The internal sources of information for innovation (*Int_sources*) represent those that arise from a firm's own departments, its staff, firms from the same group, and others. The importance of this type of information has been measured on a four-point scale (not used = 0; poor = 1; medium = 2; high = 3). Also addressed are the external sources of knowledge that a firm taps into, which are captured across a wide range of external information sources: suppliers, customers, competitors, consultants, commercial laboratories, private R&D firms, universities, technological centres, public research centres, commercial events, scientific journals and papers and professional associations. All these variables have been reduced to

Table 11.1 Variables in the study

Dependent variable	Meaning	Codification
Adoption of organizational innovation (a)	This variable counts the number of types of *organizational* during the research period: significantly improved organizational methods, new enterprise tasks organization, workplace organization or new management external relations. Each if this organizational innovation refers to the adoption of the innovation itself measured as binary variable (0–1)	Continuous (0–4)
Adoption of organizational innovation (b)	Indicates whether the enterprise has introduced at least one new or improved *organizational* innovation during the research period	Binary (0–1)
Adoption of management innovation (c)	This variable counts the number of types of *organizational* and/or marketing innovations during the research period. To the four types of organizational innovation considered in (a) this variable adds the four types of marketing innovation This dependent variable uses a broader definition of organizational innovation as Mol and Birkinshaw (2009) use including marketing innovation As types of organizational innovation, the marketing types of innovation are referred to the adoption of the innovation itself measured as binary variable (0–1)	Continuous (0–8)
Adoption of management innovation (d)	Indicates whether the enterprise has introduced at least one new or improved *organizational* innovation and/or marketing innovation during the research period	Binary (0–1)
Independent variable		
Size	Logarithm of the annual average of full-time employees in 2006	Continuous
Education of workforce	Indicates the share of employees with a high degree	Continuous
Geographic scope	Indicates the scope of locations composing a company's sales market, considering: local (1); national (2); European union, EFTA countries or EU candidates (3); and other countries (4)	Continuous

(continued)

Table 11.1 (continued)

Dependent variable	Meaning	Codification
Internal sources	The importance of the internal sources of information to innovate (by internal it is considered the firm's own departments, staff, firms from the same group, etc.) The importance of information of each source has to be in a four point scale: not used = 0; poor, value = 1; medium, value = 2; high, value = 3	0–3 interval
Industrial sources	External sources factors industry and science are the result of a PCA applied to different variables (kmo: 0.8635; % explained variance 57.53 %) corresponding to different sources of information for innovation	Continuous, from scores from the second factor analysis
Science sources	Industrial_sources: corresponds to clients, suppliers, competitors, commercial events, scientific journals and magazines, and professional associations Science_sources: corresponds to consultants, commercial laboratories, private R&D firms, universities, technological centers, and public research centers Each of the information sources refers to the importance of the information in order to innovate from each source and corresponds to the question: "in the period 2004–2006, how important have the following information sources been for the innovation activities of your enterprise?" Clients, suppliers, competitors, consultants, commercial events, scientific journals and magazines and papers, professional associations, consultants, commercial laboratories, private R&D firms, universities, technology centers, and public research centers The importance of information of each source has to be in a four point scale: not used = 0; poor, value = 1; medium, value = 2; high, value = 3	
no_Inno	Indicates whether the firm has not carried on any process or product innovations	Dummy 0–1
Inno_only_process	Indicates whether the firm has carried on only process innovations, without undertake product innovations	Dummy 0–1

(continued)

Table 11.1 (continued)

Dependent variable	Meaning	Codification
Inno_only_product	Indicates whether the firm has carried on only product innovations, without undertake process innovations	Dummy 0–1
Inno_process&product	Indicates whether the firm has carried on product and process innovations simultaneously	Dummy 0–1
Interaction terms	Following Mol and Brikinshaw (2009) and Ganter and Hecker (2012) this study computes nine interaction variables: the internal and external sources are multiplied by size, education of workforce and geographic scope: Internal × education workforce, industrial × education workforce, science × education workforce; internal × size, industrial × size, science × size; internal × geographic scope, industrial × geographic scope, science × geographic scope Ai and Norton (2003) checking is carried out	Continuous
Control variables		
Industry_NACE_code	Industry classification by NACE-93 (2-digits, 59 sectors), from 15 to 74 (baseline NACE_55)	Dummy: 0–1
Export intensity	Indicates the export intensity measuring the range of export turnover over the total turnover	Dummy 0–1
Innovation inhibitors	This variable is referred to the importance of various innovation inhibitors measured, and is calculated as a result of summating the following innovation inhibitors: At the survey each of the innovation inhibitors refers to the importance of the inhibitor in a four point scale: not effect=0; poor, value=1; medium, value=2; high, value=3. If the answer is no effect then the value of 0 is considered otherwise the value is 1	Continuous (0–11)
Public support	Indicates if the enterprise has received public financial support through local, autonomic or state administration, or the European union	Dummy: 0–1

Table 11.2 Descriptive statistics

Variable	Mean	Std. dev.	Min	Max
Organizational innovation (a)	1.255	1.394	0	4
Organizational innovation (b)	0.537	0.499	0	1
Organizational innovation (c)	1.828	2.033	0	8
Organizational innovation (d)	0.597	0.490	0	1
Size	3.728	1.263	0	11.145
Innovation inhibitors	7.710	3.318	0	11
Geographic scope	2.784	1.074	1	4
Public support	0.348	0.476	0	1
Export intensity	0.062	0.155	0	1
Education of workforce	0.229	0.260	0	1
Internal sources	2.160	1.007	0	3
Industrial sources	0.000	1.000	−2.800	3.123
Science sources	0.000	1.000	−1.497	3.964
Inno_only_process	0.359	0.480	0	1
Inno_only_product	0.168	0.374	0	1
Inno_process&product	0.341	0.474	0	1

Source: own

two factors through a PCA (KMO 0.8635 and 59 % of explained variance). The first component obtained from this PCA (*Industrial_sources*: customers, suppliers, competitors, consultants, commercial events, professional associations and commercial magazines) corresponds to sources related to industrial agents in the value chain, plus some commercial factors. The second component (*Science_sources*: commercial laboratories, private R&D firms, universities, technological centres and public research centres) corresponds to more scientific and specific pecuniary knowledge (commercial laboratories, private R&D firms, universities, technology centres and public research centres). Nevertheless, in order to match the baseline paper's results, the ex-ante search components are also tested (for the sake of brevity, results upon request).[2] Finally, in order to validate the interaction effects, we follow the same procedure as Ganter and Hecker (2012), as Ai and Norton (2003) recommend.

11.4 Results

Table 11.3 summarizes the results (Tables 11.4 and 11.5 show the full results with the interactions effects). The interaction effects are only provided in summary form below (full results available upon request). It is important to notice that the B and D

[2] Mol and Birkinshaw ex-ante classification of the search construct:

-Market sources: commercial events, scientific journals and magazines, and professional associations (Mol and Birkinsha 2009)

-Professional sources: clients, suppliers, competitors, consultants, commercial laboratories, private R&D firms, universities, technological centers, and public research centers

Table 11.3 Results from the ordered and binary logits

	A	B		C		D
	Coefficients	Coefficients	Exp (coefficients)	Coefficients	Coefficients	Exp (coefficients)
Size	.094**(.015)	0.078***	1.081***(.019)	.026*(.014)	0.0266	1.027(.018)
Innovation inhibitors	.043**(.005)	0.046***	1.048***(.006)	.045***(.005)	0.0507***	1.052***(.006)
Geographic scope	.012(.019)	0.012	1.012(.0232)	.060***(.019)	0.0559**	1.057**(.024)
Public support	–.064(.037)	–0.10**	.896**(.038)	–.096***(.037)	–0.127***	.880***(.038)
Export intensity	–0.350**(.124)	–0.36***	0.690***(.0954)	–.478***(.121)	–0.543***	.580***(.081)
Education of workforce	.670**(.086)	0.690***	1.995***(.1995)	.637***(.085)	0.6523***	1.920***(.196)
Internal sources	.137**(.018)	0.124***	1.132***(.022)	.145***(.017)	0.1274***	1.13***(.023)
Industrial sources	.306**(.018)	0.291***	1.338***(.027)	.320***(.018)	0.2850***	1.330***(.028)
Science sources	.132**(.018)	0.106***	1.112***(.023)	.110***(.017)	0.1022***	1.107***(.024)
Inno_only_process	.725**(.060)	0.803***	2.232***(.142)	.664***(.058)	0.7800***	2.18***(.137)
Inno_only_product	.204**(.068)	0.233***	1.263***(.089)	.446***(.065)	0.4853***	1.624***(.113)
Inno_process&product	1.14**(.061)	1.215***	3.372***(.222)	1.311***(.059)	1.341***	3.824***(.252)
N	12,824	12,824		12,824		12,824
LR chi2(50)	1,875.12	1,522.26		2,118.19		1,501.30
Prob>chi2	0.000	0.000		0.000		0.000
Pseudo R2	0.051	0.086		0.047		0.087
Log likelihood	–17,302.11	–8,093.45		–21,734.51		–7,893.41

Models C and D incorporates marketing activities; In each model the Industry NACE is included, with its 58 dummies (NACE 55 is the baseline). For the sake of brevity, results upon request. $^{***}p<0.01$;$^{**}p<0.05$; $^{*}p<0.1$; + Models B and D are binary logits, so the interpretation of results is not made on the coefficient, as in A and C ordered logit. In order to interpret results from the logit model we need to exponentiate the coefficients and interpret them as odds-ratio, that is, using the Exp (coefficient)

Table 11.4 Full specifications including interactions

	1(a)	2(a)	3(a)	4(a)	5(a)	6(a)	7(a)	8(a)	9(a)
Size	0.094***	0.094***	0.095***	0.020	0.094***	0.092***	0.095***	0.094***	0.093***
	(0.015)	(0.015)	(0.015)	(0.036)	(0.015)	(0.015)	(0.015)	(0.015)	(0.015)
Innovation inhibitors	0.043***	0.043***	0.043***	0.043***	0.043***	0.043***	0.043***	0.042***	0.043***
	(0.005)	(0.005)	(0.005)	(0.005)	(0.005)	(0.005)	(0.005)	(0.005)	(0.005)
Geographic scope	0.012	0.012	0.012	0.014	0.012	0.012	–0.07*	0.013	0.015
	(0.019)	(0.019)	(0.019)	(0.019)	(0.019)	(0.019)	(0.040)	(0.019)	(0.020)
Public support	–0.06*	–0.06*	–0.06*	–0.06*	–0.06*	–0.06*	–0.06*	–0.06*	–0.06*
	(0.037)	(0.037)	(0.037)	(0.037)	(0.037)	(0.037)	(0.037)	(0.037)	(0.037)
Export intensity	–0.35***	–0.35***	–0.34***	–0.35***	–0.35***	–0.35***	–0.37***	–0.34***	–0.37***
	(0.123)	(0.123)	(0.123)	(0.123)	(0.123)	(0.123)	(0.124)	(0.123)	(0.124)
Education of workforce	0.652***	0.666***	0.665***	0.674***	0.669***	0.673***	0.675***	0.667***	0.676***
	(0.194)	(0.086)	(0.086)	(0.086)	(0.086)	(0.086)	(0.086)	(0.086)	(0.086)
Internal sources	0.136***	0.138***	0.137***	0.018	0.137***	0.137***	0.036	0.137***	0.138***
	(0.022)	(0.018)	(0.018)	(0.056)	(0.018)	(0.018)	(0.045)	(0.018)	(0.018)
Industrial sources	0.306***	0.294***	0.305***	0.304***	0.305***	0.305***	0.305***	0.354***	0.306***
	(0.018)	(0.024)	(0.018)	(0.018)	(0.053)	(0.018)	(0.018)	(0.049)	(0.018)
Science sources	0.131***	0.131***	0.122***	0.130***	0.131***	0.068	0.131***	0.131***	0.012
	(0.017)	(0.017)	(0.024)	(0.017)	(0.017)	(0.049)	(0.017)	(0.017)	(0.049)
Inno_only_process	0.725***	0.723***	0.723***	0.723***	0.725***	0.723***	0.724***	0.724***	0.718***
	(0.060)	(0.060)	(0.060)	(0.060)	(0.060)	(0.060)	(0.060)	(0.060)	(0.060)
Inno_only_product	0.203***	0.201***	0.202***	0.203***	0.203***	0.200***	0.202***	0.203***	0.200***
	(0.068)	(0.068)	(0.068)	(0.068)	(0.068)	(0.068)	(0.068)	(0.068)	(0.068)
Inno_process&product	1.141***	1.139***	1.140***	1.137***	1.141***	1.137***	1.135***	1.142***	1.134***
	(0.061)	(0.061)	(0.061)	(0.061)	(0.061)	(0.061)	(0.061)	(0.061)	(0.061)
Internal × education workforce	0.007								
	(0.074)								

(continued)

Table 11.4 (continued)

	1(a)	2(a)	3(a)	4(a)	5(a)	6(a)	7(a)	8(a)	9(a)
Industrial × education workforce		0.049 (0.066)							
Science × education workforce			0.033 (0.059)						
Internal × size				0.032** (0.014)					
Industrial × size					0.001 (0.013)				
Science × size						0.016 (0.012)			
Internal × geographic scope							0.038** (0.016)		
Industrial × geographic scope								−0.01 (0.016)	
Science × geographic scope									0.040** (0.015)
Industry	Yes	Yes	Yes	Yes	Yes	Yes	Yes	Yes	Yes
N	12,824	12,824	12,824	12,824	12,824	12,824	12,824	12,824	12,824
LR chi2(50)	1,875.13	1,875.67	1,875.43	1,880.1	1,875.12	1,876.95	1,880.98	1,876.24	1,881.72
Prob>chi2	0	0	0	0	0	0	0	0	0
Pseudo R2	0.0514	0.0514	0.0514	0.0515	0.0514	0.0515	0.0516	0.0514	0.0516
Log likelihood	−17,302.10	−17,301.83	−17,301.95	−17,299.61	−17,302.16	−17,301.18	−17,299.17	−17,301.54	−17,298.804

Table 11.5 Full specifications including interactions

Exp (coefficients) in the specifications	1(b)	2(b)	3(b)	4(b)	5(b)	6(b)	7(b)	8(b)	9(b)
Size	1.080***	1.081***	1.081***	0.988	1.081***	1.078***	1.081***	1.081***	1.079***
	(0.019)	(0.019)	(0.019)	(0.039)	(0.019)	(0.019)	(0.019)	(0.019)	(0.019)
Innovation inhibitors	1.047***	1.047***	1.047***	1.047***	1.047***	1.047***	1.048***	1.047***	1.048***
	(0.006)	(0.006)	(0.006)	(0.006)	(0.006)	(0.006)	(0.006)	(0.006)	(0.006)
Geographic scope	1.012	1.012	1.012	1.013	1.012	1.012	0.939	1.011	1.016
	(0.022)	(0.022)	(0.022)	(0.022)	(0.022)	(0.022)	(0.040)	(0.022)	(0.022)
Public support	0.896**	0.895**	0.895**	0.894***	0.895**	0.893***	0.895**	0.895**	0.892***
	(0.038)	(0.038)	(0.038)	(0.038)	(0.038)	(0.038)	(0.038)	(0.038)	(0.038)
Export intensity	0.690***	0.690***	0.692***	0.682***	0.690***	0.684***	0.679***	0.692***	0.669***
	(0.095)	(0.095)	(0.095)	(0.094)	(0.095)	(0.094)	(0.094)	(0.095)	(0.092)
Education of workforce	2.196***	1.995***	1.986***	2.012***	1.994***	2.005***	2.005***	1.992***	2.012***
	(0.472)	(0.199)	(0.198)	(0.201)	(0.199)	(0.200)	(0.200)	(0.199)	(0.201)
Internal sources	1.140***	1.132***	1.132***	0.976	1.132***	1.132***	1.031	1.131***	1.132***
	(0.028)	(0.022)	(0.022)	(0.060)	(0.022)	(0.022)	(0.051)	(0.022)	(0.022)
Industrial sources	1.338***	1.336***	1.337***	1.334***	1.339***	1.337***	1.336***	1.364***	1.338***
	(0.027)	(0.035)	(0.027)	(0.027)	(0.082)	(0.027)	(0.027)	(0.076)	(0.027)
Science sources	1.112***	1.112***	1.098***	1.111***	1.112***	1.014	1.111***	1.112***	0.959
	(0.023)	(0.023)	(0.031)	(0.023)	(0.023)	(0.060)	(0.023)	(0.023)	(0.054)
Inno_only_process	2.235***	2.232***	2.227***	2.224***	2.232***	2.226***	2.226***	2.231***	2.212***
	(0.142)	(0.142)	(0.142)	(0.141)	(0.142)	(0.141)	(0.141)	(0.142)	(0.141)
Inno_only_product	1.264***	1.262***	1.261***	1.261***	1.263***	1.257***	1.260***	1.263***	1.258***
	(0.089)	(0.089)	(0.089)	(0.089)	(0.089)	(0.089)	(0.089)	(0.089)	(0.089)
Inno_process&product	3.374***	3.371***	3.368***	3.350***	3.371***	3.357***	3.350***	3.373***	3.345***
	(0.222)	(0.221)	(0.221)	(0.220)	(0.221)	(0.221)	(0.220)	(0.222)	(0.220)

(continued)

Table 11.5 (continued)

Exp (coefficients) in the specifications	1(b)	2(b)	3(b)	4(b)	5(b)	6(b)	7(b)	8(b)	9(b)
Internal × education workforce	0.959 (0.079)								
Industrial × education workforce		1.005 (0.077)							
Science × education workforce			1.044 (0.073)						
Internal × size				1.041** (0.016)					
Industrial × size					0.999 (0.015)				
Science × size						1.024* (0.015)			
Internal × geographic scope							1.036** (0.018)		
Industrial × geographic scope								0.993 (0.018)	
Science × geographic scope									1.051*** (0.019)
Industry	Yes	Yes	Yes	Yes	Yes	Yes	Yes	Yes	Yes
N	12,824	12,824	12,824	12,824	12,824	12,824	12,824	12,824	12,824
LR chi2(50)	1,522.52	1,522.27	1,522.65	1,528.54	1,522.26	1,525	1,526.38	1,522.4	1,529.88
Prob > chi2	0	0	0	0	0	0	0	0	0
Pseudo R2	0.086	0.086	0.086	0.0863	0.086	0.0861	0.0862	0.086	0.0864
Log likelihood	−8,093.326	−8,093.451	−8,093.259	−8,090.314	−8,093.453	−8,092.085	−8,091.3966	−8,093.3834	−8,089.6445

models are *logit* ones, so the figures provided are not coefficients but odds-ratios. Thus, in model B, the odds of firms adopting organizational innovation (versus not adopting) increase by a factor of 3.372. That is, a firm being a full technical (product and process simultaneously) adopter increases 3.372 times the chances of that firm being an adopter of organizational innovation. The models A and C are interpreted as usual, using the listed coefficients. The estimates underline the importance of the antecedents in both the internal resources (*context* category) – in particular: workforce, size and geographic scope – together with the relational (*search* category), made up of external sources of knowledge. Overall, the proposed framework provides good predictors of the propensity to adopt innovations in management. Our estimates fully coincide with those of Mol and Birkinshaw (except for differences mentioned below), and contradict the stated small influence of those predictors reported in Ganter and Hecker's (2012)results. Therefore, these partial similarities (with the UK), and key differences (with Germany), imply that institutional environment differences are sources of deviations, as stated in the innovation systems literature (e.g. Furman et al. 2002).

In particular, the *size* variable is positive and significant, confirming a body of literature which has emphasized size as an important driver to explain inducements to technical innovation (Cohen and Klepper 1996; Damanpour 2010; Klepper 1996; Nord and Tucker 1987; Reichstein and Salter 2006). Similarly, management innovation studies have also predicted and confirmed a positive sign (Ganter and Hecker 2012; Mol and Birkinshaw 2009). The positive result evidenced in the impact of the quality of education of the workforce on the adoption of management innovation, (education workforce variable), also confirms the results of previous studies (Alvesson 1995; Chandler 1962; Daft 1978; Mol and Birkinshaw 2009). Also, external knowledge sources, in general, are found to be drivers of the innovation process in firms, as found in other studies (Barge-Gil 2010; Cabagnols 1999; Escribano et al. 2009; Reichstein and Salter 2006; Rouvinen 2002; Vega-Jurado et al. 2008; Von Hippel 1988), and specifically in those works which recognize the value of both active search of external sources (e.g. Birkinshaw et al. 2008; Mol and Birkinshaw 2009) and internal sources (e.g. Hansen and Løvås 2004) for the introduction of new management innovation activities. However, our results contradict the small influence (only professional sources is positive and significant) on the adoption of management innovation found in Ganter and Hecker (2012).

There are, however, a few minor nuances to be remarked upon. In relation to the German case in Ganter and Hecker (2012), we find dissimilar results for geographic scope and similar ones for public support. Thus, in our work, the *geographic scope* variable is positive and significant in all models, showing that exposure to international environments by Spanish firms fosters the adoption of management innovation, which are different results to those of Ganter and Hecker. The control variable *public support* is not an antecedent of management innovation adoption in Spain, as in the German and UK cases. Also, *export* intensity is negative and significant in all models, fact similar to the UK case. Innovation inhibitors are also positive and significant, as in the UK and German cases. Finally, it is clear that industry differences matter when it comes to explaining the adoption of management innovation, matching the baseline results.

Table 11.6 External sources ex-post using PCA (industrial and science)

Interactions		B: inno_org (0–1)	A: inno_org_sum (0–4)	D: inno_manag (0–1)	C: inno_manag_sum (0–8)
Internal_sources	Education Workforce	no	No	no	no
Industrial_sources	Education Workforce	no	No	no	no
Science_sources	Education Workforce	no	No	no	no
Internal_sources	Size	$+^{**}$	$+^{**}$	$+^{***}$	$+^{**}$
Industrial_sources	Size	$+^{***}$	No	no	no
Science_sources	Size	no	No	$+^{*}$	no
Internal_sources	Geographical scope	$+^{*}$	$+^{**}$	$+^{**}$	$+^{**}$
Industrial_sources	Geographical scope	no	No	no	no
Science_sources	Geographical scope	$+^{**}$	$+^{**}$	$+^{***}$	$+^{**}$

$^{***}p(0.01)$; $^{**}p(0.05)$; $^{*}p(0.1)$

Table 11.7 External sources following Mol and Birkinshaw (market and professional)

Interactions		B: inno_org (0–1)	A: inno_org_sum (0–4)	D: inno_manag (0–1)	C: inno_manag_sum (0–8)
Internal	Education Workforce	no	no	no	no
Market	Education Workforce	no	$+^{*}$	no	$+^{*}$
Professional	Education Workforce	no	no	no	no
Internal	Size	$+^{**}$	$+^{**}$	$+^{***}$	$+^{**}$
Market	Size	no	no	$+^{*}$	no
Professional	Size	no	no	no	no
Internal	Geographical scope	$+^{**}$	$+^{**}$	$+^{**}$	$+^{***}$
Market	Geographical scope	$+^{***}$	$+^{**}$	$+^{***}$	$+^{***}$
Professional	Geographical scope	no	no	no	no

$^{***}p(0.01)$; $^{**}p(0.05)$; $^{*}p(0.1)$

There are key differences in respect of hypothesis 5a, 5b, and 6. Tables 11.6 and 11.7 show a summary of the interactions representing the fifth hypothesis. As a robustness check, we have run the interactions using the *ex-ante* classification of external sources of knowledge (market and professional sources, following Mol and Birkinshaw). Results using the ex-ante grouping are fairly similar to the ones obtained from doing the ex-post (PCA) classification, although all of them are different from those shown in the baseline papers. First, our results fully contradict those of Mol and Birkinshaw. Indeed, we find a positive (complementary) relationship, which is translated into a higher necessity to tap into, and have access to, external sources of knowledge when firms increase in size, workforce education and geographic scope. Therefore, the greater the internal (context) resources, the more there is access to relational (search) knowledge, confirming the *absorptive capacity* construct (Cohen and Levinthal 1990) as a key organizational learning driver. In fact, the latter predicts the adoption of new management practices. Our results also contradict those interactions in Ganter and Hecker (2012) which show no effect on the adoption of the management mode. These results are extremely important, since in Mol and Birkinshaw (2009) the interactions show a substitution (negative) or mitigation effect for adopting new management practices. On the contrary, in our paper, the interaction terms, in respect of internal (contextual) capabilities and relational (search) ones, are in most cases positive and significant, except for those relating to the Education Workforce variable. As a matter of fact, *size X Internal_sources* is positive and significant (at $p<0.05$ and $p<0.01$) in all models; *size X Industrial_sources* is positive and significant in model B. Then, geographical scope is positive and significant when multiplied by internal and science sources, nor in the industrial sources. In respect of Mol and Birkinshaw's search approach, our results agree with our ex-post results in the variables *size X market sources* and *geographical scope X market sources*. The rest of the interactions, specifically those referring to the highly educated workforce of the firm (Education Workforce), show no significant results, except in the case of *market sources* X *Education Workforce*. All in all, the interactions terms which are positive and significant show an *amplifying* effect, that is, the greater the geographical scope of the market in which the firm is operating, the greater the internal sources of knowledge the firm has access to, due to the existence of more information sources from its internal operations abroad, and thus the higher the probability to engage in the introduction of new management activities. Similarly, the larger the firm, the higher the level of access to, and the use of, internal sources of knowledge and industrial sources, thus increasing the likelihood of new adoption of organizational innovation. In short, our results show a complementary (rather than substitution) effect, meaning that the combination of internal and external capabilities (or contextual and search, in Mol and Birkinshaw terms) increases the likelihood of new management activities adoption, complementing each other. This result is similar to that of Escribano et al. (2009), and of Cassiman and Veugelers (2006), although those works related to technological innovation adoption, rather than to management innovation. See Tables 11.6 and 11.7.

Regarding the extension of the model depicted in hypothesis 6, it is observed how product and process innovations are also antecedents of management innovation adoption, that is, they exhibit *complementarities* (Milgrom and Roberts 1995) when there is synchronous adoption (Ettlie 1988). This result is similar to those of the baseline papers, although those papers use product and process as control variables while we show the existence of complementarities by using a binary variable which reflects the fact that the company synchronously adopts all forms of technological innovation (product and process simultaneously, as full technical innovators). In fact, the latter variable has the highest coefficient for all models and variables, and confirms the socio-technical view, and the synchronous adoption previously depicted (Inno_process&product 1.14, $p<0.01$, process adoption 0.72 $p<0.01$ and product 0.20 at $p<0.01$, in model A). Put differently, when a firm is a complex innovator co-adopting all technological innovations (product and process simultaneously), then the coefficient of this technical co-adoption is a good predictor of the propensity to adopt management innovations. Why? Basically there are two reasons. Firstly, being a full-technical innovator (products and process simultaneously) requires management innovation in order to couple technology with the organization(e.g. Fleck 1994; Leonard-Barton and Deschamps 1988). This idea is reflected in the concept *organizational integration* (Ettlie and Reza 1992), or joint optimization of practices that are socially and technically-oriented (Cua et al. 2001; Damanpour et al. 2009). Secondly, complementarities from that integration are most likely to materialize in a system of complex interactions among multiple elements. Thus, complementarities are effective when embedded in an overall system which involves many elements(Ennen and Richter 2010: 224), involving technology and organization. This fact is also recognized by Cassiman and Veugelers (2006) who argue that success in innovation requires combining not only the right innovation practices, but also creating an organizational context in which these practices enhance one other. In short, our results confirm Lam's (2005) statement that *organizational innovation* is a precondition for ensuring innovation in organizations and that it is necessary to study the relevant and key organizational characteristics which enhance a firm's capacity for innovation(e.g. Hall 1992, 1993; Henderson and Cockburn 1994). In Table 11.8, a summary of our hypotheses, and their acceptance or rejection in the revisited models, is shown. See Table 11.8.

In Tables 11.9 and 11.10 we show additional information on the effect of synchronous adoption. As shown in Table 11.9 (calculations in columns), the full technology innovators (product and process innovation simultaneously) which do not adopt new organizational methods (1,420; 24 % of the sample) are outnumbered by full technology innovators which do introduce new organizational practices (i.e., complex innovators 2,943; 43 % of the sample). Similarly, the full technology innovators which do not introduce new management (including marketing) activities (1,137; 22 %) are also outnumbered by the full technology innovators which do introduce new management activities (again, complex innovators, 3,232; 42 %). These results confirm those of Schmidt and Rammer (2007) for Germany (see Sect. 11.1), using CIS data, in which only a minority of firms introduce solely organizational (573, 8 % of our sample) or management innovations (649, 8 % of our sample), compared to the majority of firms which do that simultaneously with technological innovations (organization methods, 43 % and management methods, 42 %). In Table 11.10, we provide similar results. The full innovators who introduce organizational innovations (model B, 2,947; 67 % of the sample) outnumber the ones which do not (model B, 1,420; 33 % of the sample); and the picture is similar for the introduction of new management innovations (model D, 3,232; 74 % of the sample versus, 1,137 representing 26 % of the sample). Overall, the interpretation of Tables 11.9 and 11.10 supports hypothesis 6 concerning synchronous adoption. Therefore, it can be pointed out that the adoption of management innovation is frequently observed in tandem with the introduction of technological innovation, corroborating Ettlie (1988), Ettlie and Reza (1992), Damanpour and Evans (1984) and Fleck (1994), among others. See Tables 11.9 and 11.10.

Finally, our paper provides a comprehensive framework for positioning management innovation, complementing that of Mol and Birkinshaw. According to the resource-based view, the distinctiveness of a firm's capabilities depends on their *rarity, inimitability, non-substitutability* and *value* (Amit and Schoemaker 1993; Barney 1991; Wernerfelt 1984). Our results are in line with the resource-based view

Table 11.8 Hypotheses and results from the UK, German and Spanish cases

Hypothesis	Our study	Mol and Birkinshaw	Ganter and Hecker
Hypothesis 1: The larger the firm, the higher the level of adoption of new organizational innovations	Confirmed	Confirmed	Confirmed
Hypothesis 2: The more highly educated the workforce of the firm, the higher the level of introduction of new management practices	Confirmed	Confirmed	Confirmed
Hypothesis 3: The greater the geographical scope of the market the firm is operating in, the higher the level of introduction of new management practices	Confirmed	Confirmed	Rejected
Hypothesis 4a: The more internal sources the firm interacts with, the higher the level of introduction of new management practices	Confirmed	Confirmed	Rejected
Hypothesis 4b: The more external sources of knowledge the firm interacts with, the higher the level of introduction of new management practices	Confirmed	Confirmed	Partially confirmed (professional or science sources)
Hypothesis 5a: The effect of internal sources on the introduction of new management practices is mitigated by size, education of the workforce, and geographic scope of the firm	Rejected Complementary effect evidenced	Confirmed	Rejected
Hypothesis 5b: The effect of external sources of knowledge on the introduction of new management practices is mitigated by size, education of the workforce, and geographic scope of the firm	Rejected Complementary effect evidenced	Confirmed	Rejected
Hypothesis 6: The introduction of new technological (product and process) innovations fosters the adoption of management innovation due to the required integration of technology in the organization	Confirmed	Partially confirmed	Partially confirmed

Source: own

Table 11.9 Analysis of synchronous adoption using the B and D models (binary)

	Whole sample N=12,824	Organization B model, value=0 N=5,163	Organization B model, value=1 N=7,661	Management D model, value=0 N=5,163	Management D model, value=1 N=7,661
no_Inno	1,687(13 %)	1,114 (19 %)	573 (8 %)	1,038(20 %)	649(8 %)
Inno_only_process	4,608(36 %)	2,155(36 %)	2,453(36 %)	1,966(38 %)	2,642(34 %)
Inno_only_product	2,160(17 %)	1,254 (21 %)	906(13 %)	1,022(20 %)	1,138(15 %)
Inno_process&product	4,369(34 %)	1,420 (24 %)	2,949(43 %)	1,137(22 %)	3,232(42 %)
Total	100 %	100 %	100 %	100 %	100 %

Source: own, calculations in columns, management in the D model includes both organizational and marketing

Table 11.10 Analysis of synchronous adoption using the B and D models (binary)

	Whole sample N=12,824	Organization B model, value=0 N=5,163	Organization B model, value=1 N=7,661	Total	Management D model, value=0 N=5,163	Management D model, value=1 N=7,661
No innovation in product or process	1,687	1114(66 %)	573(34 %)	100 %	1,038(62 %)	649(38 %)
Inno_only_process	4,608	2,155(47 %)	2,453(53 %)	100 %	1,966(43 %)	2,642(57 %)
Inno_only_product	2,160	1,254(58 %)	906(42 %)	100 %	1,022(47 %)	1,138(53 %)
Inno_process&product	4,369	1,420 (33 %)	2,949(67 %)	100 %	1,137(26 %)	3,232(74 %)

Source, own; calculations in rows, management in the D model includes both organizational and marketing of the firm, in the sense that complex interrelationships among the technological and management innovations generate difficult to imitate strategic assets which prevent imitation due to the complementary and ambiguity of those systems (e.g. Rivkin 2000). Similarly, this is what Teece (1986) suggest about *complementary assets*.

11.5 Conclusions

This paper explores the drivers or antecedents of the decision to introduce new management activities, that is, management innovation adoption, elaborating on the Mol and Birkinshaw (2009) and Ganter and Hecker (2012) baseline models. This paper covers 12,824 firms from 2006 CIS data in Spain, and expands on those models by introducing the idea of synchronous co-adoption using a resource-based view. According to our results, those models are reassessed and confirmed, although in the Spanish results some key differences stand out. All in all, most hypotheses are confirmed, except those addressing interactions. Confronting our results with those in the baseline papers, this paper's findings resonate quite well, except for the results regarding geographical scope and external (search) sources (for the German case), export intensity (for the German cases), and interaction effects (for both the UK and Germany). In addition, the extended hypothesis about synchronous adoption is accepted, going a step further on the relationship between technology and management innovation.

In general, the divergent outcomes in management adoption behaviour between the Spanish and the UK and German cases represent differences in the underlying institutional environments. In this sense, we fully agree with the assertion that there are country-specific paths of innovation adoption, as Ganter and Hecker (2012: 7) suggest. We contradict the suggestion made by Ganter and Hecker (2012: 7) that "For German firms, the result suggests that organizational innovation is an important driver of product and process innovation and, thus a source of competitive advantage". In fact, that work does not provide evidence of a sequential path of cause-and-effect adoption, in which organizational innovation drives product and process innovation. Conversely, we present results which confirm a synchronous co-adoption, rather than a sequential path. Evidence of a sequential path will require the use of panel data or longitudinal studies. Our paper provides evidence that the relationship between organizational subsystems is a correlative relationship representing a "coupling of dissimilarities", whereby each change in a subsystem requires alterations in the other subsystems (Trist and Murray 1993). Following the results from the last hypothesis about complementarities, the main message outlined in this paper is that it is necessary to encourage firms to adopt a more comprehensive and systemic view of the innovation process, integrating new technology (product and process innovations) in the organization by introducing new management practices. Therefore, according to our evidence, it can be pointed out that the adoption of management innovation is usually observed in tandem with that of technological innovation, as suggested by the literature (e.g. Ettlie 1988).

All in all, in order to go beyond the divergent results in innovative firm behaviour between the three countries, this study attempts to build a more comprehensive picture of the drivers influencing organizational innovation adoption using a resource-based view of the firm in order to complement the context and search constructs first used in Mol and Birkinshaw (2009). Thus, scholars will now have a more comprehensive model for predicting the adoption of management innovation, and thus the innovation process will be better understood beyond the classic technological (product and process) imperative.

This paper has contributed to the literature in the following ways. Firstly, the paper provides insights into the drivers of management innovation adoption, extending the emerging empirical literature on management innovation. Secondly, the paper also contributes to the management innovation literature by exploring its complementary role with the technological mode, analysing the complementary effects arising from simultaneous co-adoption (Damanpour and Evan 1984), synchronous co-adoption (Ettlie 1988) or organizational integration (Ettlie and Reza 1992). The paper also contributes by formulating a better theoretical connection between technological innovation and management innovation, offering new lenses for a broad tackling of the organizational innovation construct. Therefore, the *conversation* about organizational innovation can be improved by addressing the complementary effects obtained from the intersection of an organization's social and technical systems, using a resource-based view of the firm.

Finally, this paper has similar limitations to the baseline papers, especially regarding the nature of the CIS data (testing for causality), the limited availability of data describing the UK and German institutional environments, and the lack of proper panels to conduct a longitudinal study. For future research, more countries need to be researched in order to unfold the research gaps in our knowledge of management innovation adoption, together with specific questionnaires which provide a more complete picture of organizational innovation strategy.

Acknowledgments This work was supported by the ECO2010: 17318 Innoclusters Project funded by the Spanish Ministry of Economics and Competitiveness. Usual disclaimers apply. Dr. Hervas-Oliver is also in debt to the Generalitat Valenciana for the visiting funding to London School of Economics, LSE (UK) from the BEST 2011 program. Furthermore, Authors gratefully acknowledge support from the Universitat Politècnica de València through the project Paid-06-12 (Sp 20120792).

References

Abrahamson E (1996) Management fashion. Acad Manage Rev 21:254–285

Ai C, Norton EC (2003) Interaction terms in logit and probit models. Econ Lett 80(1):123–129

Aiken M, Hage J (1971) The organic organization and innovation. Sociology 5(1):63

Alvesson M (1995) Management of knowledge-intensive companies. Walter de Gruyter Inc., Berlin/New York

Amit R, Schoemaker PJH (1993) Strategic assets and organizational rent. Strategic Manage J 14(1):33–46

Armbruster H, Bikfalvi A, Kinkel S, Lay G (2008) Organizational innovation: the challenge of measuring non-technical innovation in large-scale surveys. Technovation 28(10):644–657
Atuahene-Gima K (1995) An exploratory analysis of the impact of market orientation on new product performance. J Prod Innov Manage 12(4):275–293
Barge-Gil A (2010) Cooperation-based innovators and peripheral cooperators: an empirical analysis of their characteristics and behavior. Technovation 30(3):195–206
Barney J (1991) Firm resources and sustained competitive advantage. J Manage 17(1):99–120
Bessant J (2003) Challenges in innovation management. In: International handbook on innovation. Elsevier Science Ltd, Oxford, pp 761–774
Birkinshaw J, Hamel G, Mol M (2008) Management innovation. Acad Manage Rev 33(4): 825–845
Cabagnols A (1999) The determinants of the continuity and consistency of the innovative behaviour of product and process innovators. European meeting on applied evolutionary economics,Grenoble, 7–9 June 1999
Cassiman B, Veugelers R (2006) In search of complementarity in innovation strategy: internal R&D and external knowledge acquisition. Manage Sci 52(1):68–82
Cebon P, Newton P, Noble P (1999) Innovation in firms: towards a framework for indicator development. Melbourne business school, working paper
Chandler AD (1962) Strategy and structure: chapters in the history of the American industrial enterprise. MIT Press, Cambridge
Cohen W, Klepper S (1996) Firm size and the nature of innovation within industries: the case of process and product R&D. Rev Econ Stat 78(2):232–243
Cohen W, Levinthal D (1990) Absorptive capacity: a new perspective on learning and innovation. Adm Sci Q 35(1):128–152
Coombs R, Hull R (1998) Knowledge management practices and path-dependency in innovation. Res Pol 27(3):239–256
Crepon B, Duguet E, Mairesse J (1998) Research, innovation, and productivity: an econometric analysis at the firm level. Econ Innov New Technol 7(2):115–159
Crossan M, Apaydin M (2010) A multi-dimensional framework of organizational innovation: a systematic review of the literature. J Manage Stud 47(6):1154–1191
Cua KO, McKone KE, Schroeder RG (2001) Relationships between implementation of TQM, JIT, and TPM and manufacturing performance. J Oper Manage 19(6):675–694
Cummings TG, Srivastva S (1977) Management of work: a socio-technical systems approach. Kent: comparative administration research institute : distributed by Kent State University Press. Translation of: (Kent, Ohio)
Daft RL (1978) A dual-core model of organizational innovation. Acad Manage J 21(2):193–210
Damanpour F (1987) The adoption of technological, administrative, and ancillary innovations: impact of organizational factors. J Manage 13(4):675
Damanpour F (2010) An integration of research findings of effects of firm size and market competition on product and process innovations. Brit J Manage 21(4):996–1010
Damanpour F, Aravind D (2011) Managerial innovation: conceptions, processes, and antecedents. Manage Organ Rev 8(2):423–454
Damanpour F, Evan WM (1984) Organizational innovation and performance: the problem of organizational lag. Adm Sci Q 29(3):392–409
Damanpour F, Walker RM, Avellaneda CN (2009) Combinative effects of innovation types and organizational performance: a longitudinal study of service organizations. J Manage Stud 46(4):650–675
Dodgson M (1993) Technological collaboration in industry: strategy, policy, and internationalization in innovation. Routledge, London/New York
Dyer JH, Singh H (1998) The relational view: cooperative strategy and sources of interorganizational competitive advantage. Acad Manage Rev 23(4):660–679
Ennen E, Richter A (2010) The whole is more than the sum of its parts or is it? A review of the empirical literature on complementarities in organizations. J Manage 36(1):207–233
Escribano A, Fosfuri A, Tribo JA (2009) Managing external knowledge flows: the moderating role of absorptive capacity. Res Pol 38(1):96–105

Ettlie JE (1988) Taking charge of manufacturing: how companies are combining technological and organizational innovations to compete. Jossey-Bass, San Francisco, p 643
Ettlie JE, Reza EM (1992) Organizational integration and process innovation. Acad Manage J 35(4):795–827
Evangelista R, Vezzani A (2010) The economic impact of technological and organizational innovations a firm-level analysis. Res Pol 39(10):1253–1263
Fleck J (1994) Learning by trying: the implementation of configurational technology. Res Pol 23(6):637–652
Furman JL, Porter ME, Stern S (2002) The determinants of national innovative capacity. Res Pol 31(6):899–933
Ganter A, Hecker A (2012) Deciphering antecedents of organizational innovation. J Bus Res 66(5):575–584
Garvin DA (1993) Building a learning organization. Harv Bus Rev 71(4):78–91
Grant RM (1996a) Prospering in dynamically-competitive environments: organizational capability as knowledge integration. Organ Sci 7(4):375–387
Grant RM (1996b) Toward a knowledge-based theory of the firm. Strategic Manage J 17:109–122
Guler I, Guillén MF, Macpherson JM (2002) Global competition, institutions, and the diffusion of organizational practices: the international spread of ISO 9000 quality certificates. Adm Sci Q 47(2):207–232
Hall R (1992) The strategic analysis of intangible resources. Strategic Manage J 13(2):135–144
Hall R (1993) A framework linking intangible resources and capabilities to sustainable competitive advantage. Strategic Manage J 14(8):607–618
Hansen MT, Løvås B (2004) How do multinational companies leverage technological competencies? Moving from single to interdependent explanations. Strategic Manage J 25(8-9):801–822
Helfat CE, Finkelstein S, Mitchell W (2007) Dynamic capabilities: understanding strategic change in organizations. Wiley-Blackwell, Malden
Henderson R, Cockburn I (1994) Measuring competence? Exploring firm effects in pharmaceutical research. Strategic Manage J 15(S1):63–84
Ichniowski C, Shaw K, Prennushi G (1997) The effects of human resource practices on manufacturing performance: a study of steel finishing lines. Am Econ Rev 87(3):291–313
Kimberly JR, Evanisko MJ (1981) Organizational innovation: the influence of individual, organizational, and contextual factors on hospital adoption of technological and administrative innovations. Acad Manage J 24:689–713
Klepper S (1996) Entry, exit, growth, and innovation over the product life cycle. Am Econ Rev 86(3):562–583
Kogut B, Parkinson D (1993) The diffusion of American organizing principles to Europe. In: Country competitiveness: technology and the organization of work. Oxford university press, New York, pp 179–202
Kogut B, Zander U (1992) Knowledge of the firm, combinative capabilities, and the replication of technology. Organ Sci 24(4):383–397
Lam A (2004) Organizational innovation. Brunel Business School, Brunel
Lam A (2005) Organizational innovation. In: Fagerberg J, Mowery DC, Nelson RR (eds) The oxford handbook of innovation. Oxford, Oxford University Press, pp 115–147
Laursen K, Salter A (2006) Open for innovation: the role of openness in explaining innovation performance among UK manufacturing firms. Strategic Manage J 27(2):131–150
Lee YS (1996) Technology transfer and the research university: a search for the boundaries of university-industry collaboration. Res Pol 25(6):843–863
Lee C, Lee K, Pennings JM (2001) Internal capabilities, external networks, and performance: a study on technology-based ventures. Strategic Manage J 22(6-7):615–640
Leonard-Barton D, Deschamps I (1988) Managerial influence in the implementation of new technology. Manage Sci 34(10):1252–1265
McEvily B, Zaheer A (1999) Bridging ties: a source of firm heterogeneity in competitive capabilities. Strategic Manage J 20(12):1133–1156

Milgrom P, Roberts J (1995) Complementarities and fit strategy, structure, and organizational change in manufacturing. J Account Econ 19(2):179–208
Mol M, Birkinshaw J (2009) The sources of management innovation: when firms introduce new management practices. J Bus Res 62(12):1269–1280
Nahapiet J, Ghoshal S (1998) Social capital, intellectual capital, and the organizational advantage. Acad Manage Rev 23(2):242–266
Nelson R, Winter S (1982) An evolutionary theory of economic change. Harvard University Press, editor. Belknap press, Massachusetts/London
Nonaka I, Takeuchi H (1995) The knowledge-creating company: how Japanese companies create the dynamics of innovation. Oxford University Press, USA
Nord W, Tucker S (1987) Implementing routine and radical innovations. Lexington books, New York
OECD (2005) Oslo manual: guidelines for collecting and interpreting innovation data. OECD, Paris
Pasmore W, Francis C, Haldeman J, Shani A (1982) Sociotechnical systems: a North American reflection on empirical studies of the seventies. Hum Relat 35(12):1179–1204
Powell WW, Koput KW, Smith-Doerr L (1996) Interorganizational collaboration and the locus of innovation: networks of learning in biotechnology. Adm Sci Q 41(1):116–145
Reichstein T, Salter A (2006) Investigating the sources of process innovation among UK manufacturing firms. Ind Corp Change 15(4):653–682
Rivkin JW (2000) Imitation of complex strategies. Manage Sci 46(6):824–844
Ross PF (1974) Innovation adoption by organizations. Pers Psych 27(1):21–47
Rouvinen P (2002) Characteristics of product and process innovators: some evidence from the Finnish innovation survey. Appl Econ Lett 9(9):575–580
Rowley T, Behrens D, Krackhardt D (2000) Redundant governance structures: an analysis of structural and relational embeddedness in the steel and semiconductor industries. Strategic Manage J 21(3):369–386
Schmidt T, Rammer C (2007) Non-technological and technological innovation: strange bedfellows? ZEW discussion papers 07-052
Singh J (2005) Collaborative networks as determinants of knowledge diffusion patterns. Manage Sci 51(5):756–770
Staw BM, Epstein LD (2000) What bandwagons bring: effects of popular management techniques on corporate performance, reputation, and CEO pay. Adm Sci Q 45(3):523–556
Teece DJ (1986) Profiting from technological innovation: implications for integration, collaboration, licensing and public policy. Res Pol 15(6):285–305
Teece D, Pisano G, Shuen A (1997) Dynamic capabilities and strategic management. Strategic Manage J 18(7):509–533
Tether BS (2002) Who co-operates for innovation, and why: an empirical analysis. Res Pol 31(6):947–967
Trist EL (1978) On socio-technical systems. In: Sociotechnical systems: a sourcebook. University Associates, San Diego
Trist EL, Bamforth KW (1951) Some social and psychological consequences of the longwall method: an examination of psychological situation and defences of a work group in relation to the social structure and technological content of the work system. Hum Relat 4(1):3–38
Trist E, Murray H (1993) The socio-technical perspective. University of Pennsylvania Press, Pennsylvania
Vaccaro IG, Jansen JJP, Van Den Bosch FAJ, Volberda HW (2012) Management innovation and leadership: the moderating role of organizational size. J Manage Stud 49(1):28–51
Vega-Jurado J, Gutiérrez-Gracia A, Fernández-de-Lucio I, Manjarrés-Henríquez L (2008) The effect of external and internal factors on firm's product innovation. Res Pol 37(4):616–632
Von Hippel E (1988) The sources of innovation. Oxford University Press, New York
Walker RM, Damanpour F, Devece CA (2011) Management innovation and organizational performance: the mediating effect of performance management. J Public Admin Res Theor 21(2):367
Wernerfelt B (1984) A resource-based view of the firm. Strategic Manage J 5(2):171–180

Index

J.-L. Hervás-Oliver and M. Peris-Ortiz (eds.), *Management Innovation*,
Springer Proceedings in Business and Economics, DOI 10.1007/978-3-319-03134-7,
© Springer International Publishing Switzerland 2014